THE FLOWER YARD

growing flamboyant flowers in containers

written and photographed by

ARTHUR PARKINSON

For my mum, Jill

'He went into the conservatory and looked out through the tinted glass, but
through the vivid colours of the window pane, it wasn't a garden that he saw,
but a tropical landscape with tigers and panthers burning in the shrubberies,
and blue parrots screaming soundlessly in the trees.'

Forrest Reid, *Apostate*

KYLE BOOKS

contents

INTRODUCTION

This book is an invitation and somehow, I hope, an education on how to garden in a truly small space. It is a calling out against what is small-garden dysmorphia, where gardens with lawns, sheds and even greenhouses are indeed called small.

The chapters follow a year of growing to create specific displays of plants – one each for spring and for summer – for a flamboyant and defiant show in pots; it is about the plants themselves and how these alone have the power to transform daily life into the beautiful.

A good friend once called my garden a path of pots; an apt title, really, for a space that measures 5 metres (16 feet) long and is filled cheek by jowl with containers on either side, leading from the pavement to our front door.

Make your world alive; even a window box has the power to unleash vitality for the greater good of your mind and of the planet, a nurturing of flora and fauna alike, creating visual and mental sanctuaries.

The yard

Where we live once had an uneven dirt surface of sandy gravel resembling a dry riverbed that formed great pools when it rained, creating a paradise for dabbling in Wellington boots for my younger brother and me. Our little cottage is the first of several in a line, side by side with a façade of sandstone – the row was once a malthouse. Derelict buildings used to tuck us away from the surround of the town, despite us being slap bang in its middle. We could not be seen from the road. These are now gone, knocked down – no bad thing in the name of redevelopment but in this swoop, a magnificent giant of a copper beech tree was felled too, opening up a vast sky and revealing us suddenly to the neighbouring town. The resulting sun has been the only reward, and all the plants mentioned in this book relish its rays.

An endlessly roaring new road that allows traffic to speed along, out into the ever-more-packed heart of a connecting, rapidly engulfing city landscape, never stops. Our town has almost lost its green belt now. Gone within my lifetime are its inner town playing fields and the farmland outskirts, as more and more dolls' houses relentlessly pop up.

The romantic farmyard surface of the yard is no more. It has been replaced with a smooth, uniform, health-and-safety-clipboard-ticked-off canvas of porous cubes. Safe and tamed, no more scuffed knees, no more puddles, dull, flat and human. And so it is the garden that has been the saviour in what would otherwise have been an irreversibly changed place to live. It offers, right by the door, an exciting stamp of colour and vibrancy that, vitally, brings life. This in turn gives me hope, essential hope, within what is now an Earth being smothered by us, humanity. We may have forgotten what Eden is but through gardening this connection can be made strong again.

The garden is central to my life; it is a daily therapy. Its seasonal highlights and growing calendar fill my head with the excitements and longings of its emerging and temporary beauty. I am constantly thinking about the next season. This growing of flowers from spring to autumn is an ever-changing, living ballet that adds mental vigour through the year. It requires months of planning, patience and constant care. You can never be truly head stuck if you immerse your mind in the needs of a flower garden. This is deliberately not low-maintenance gardening but a time-requiring, orchestrated, beautiful, admittedly expensive obsession and addiction. I like the challenge of growing a floral jungle to block out the rawness of what urban living often is – an existence devoid of life, sterile and detached.

The realities for my generation, in not having enough money to garden, involve situations such as mine, where you may still live with parents or rent. Gardening in pots allows a garden to be transported from one situation to another. We may not be able to put down our roots as readily as plants do, but containers mean anyone can grow a garden that can be moved easily.

To be a gardener is to be an artist; you are painting a living picture, one that is never finished but that continues to evolve, delight and draw on our mind's imagination. There seems to be an abhorrent idea that beauty is something you can only aspire to and achieve in a grand house or in a large garden. But beauty doesn't bow to this; it can be nurtured in the smallest space, and here it can be truly fabulous. Perhaps, best of all, is its potential to connect and support the natural world around us, a world that is just waiting to be given the vital help to thrive once more.

VISIONS AND DREAMS

My ultimate dream, in terms of space, would be to have a modest half-dirt, half-cobbled brick courtyard surrounded by partly converted stables. It is good to dream big, I think, and then to convert elements of this imagination into reality, somehow.

Figs would be espaliered across the courtyard's loose mortar walls. A room-like setting, reminiscent of the Dutch painter Melchior d'Hondecoeter's scenes of fowls and flowers. *The Best Exotic Marigold Hotel* meets *My Family and Other Animals*; dilapidated but romantic, a floral aviary really, ornate cages and terracotta pots with tortoises.

Best of all, though, this illusion of grandeur would be completely detached: no neighbours, no traffic noise, no plastic dustbins. What a privilege it must be to be able to stagger out drunkenly into the night, half-naked, singing 'MacArthur Park' to oneself amongst the flowers, without worrying that the neighbours might call the police.

I am not, however, desperate for a larger garden. I find the challenge of conquering the restrictions of an urban environment hugely thrilling. I love small town gardens, by which I mean gardens where plants come first in abundance. I have no desire for endless herbaceous borders, which so easily become tired and full of perennial weeds. Give me a flock of dolly tubs any day, ideally on old bricks or York stone. An old orchard would be, admittedly, heaven though, for hens.

So many urban gardens are given up to paving, decking, Astro turf, gravel, bark chippings, barbecues – even plastic flowers. We are often told that UK gardens collectively cover some one million hectares – more than the combined area of national nature reserves. But I shudder to imagine how many hectares consist of hard standing rather than cultivated soil. I walk past hundreds of these shells each day, coffin gardens!

LEFT An avenue of beetroot purple hyacinth 'Woodstock', and hyacinths 'Anastasia' and 'Kronos' and *Muscari latifolium* fill the air with perfume. All of these bulbs are perennial in habit and provide nectar for pollinators in early spring.

OVERLEAF Viola 'Tiger Eye Red'. Having arrived as seedling plugs a few weeks previously, they are grown on for a while in these old terracotta pots where they look striking, gaggled together on an old bistro table outside. They will all be planted into larger pots before the winter as they would quickly outgrow and then go to seed in these.

ARCHIPELAGOS OF GALVANISED METAL & TERRACOTTA

I garden in pots because I do not have a choice, but I rarely resent this as it is like having great living vases of growing flower arrangements. You can fill pots easily, cramming them with colour and textures, creating islands of flamboyance.

Good, big pots elevate the garden helpfully for a small space. This is wonderful as you get floral grandeur that is uplifted to almost hip height immediately. Plants such as violas that you would otherwise have to get on your knees to properly study require just a light bow to admire. If you place large pots around seating, a tulip behind you can seemingly be perched on your shoulder like a parrot. Summer flowers can be literally in, around or above your face, towering and engulfing.

Pot gardening is fast and satisfying because the whole job of planting and replanting is done very quickly, a sharp look for a whole season can be prepared within an afternoon. You can also treat a container garden like a changing stage, with the pots as props that can be moved around as you see fit. Compare this to having to maintain a constant expanse of earth that at times can be either a great mud slick or baked hard, requiring a good dose of backache to get it looking ship-shape.

The right size pots

Big is essential for scale and ease of care. Large and heavy pots are more resistant to storms, too, and this is worth remembering in our increasingly hostile climate. For summer especially, remember that a pot needs to act as an anchor for something that is going to grow tall and be rocked about in the wind; if it is not heavy or broad enough, the whole thing will topple over.

One summer, we got hit horrendously by flash flooding; it was an almost Biblical flood. The yard was deep in water, as was the high street. It was a strange scene, and when such a thing happens you are a mere parasite scrambling about.

Thankfully, though, our garden was not lost; the large pots all stood firmly against and above this flash tide and as the waters calmed and drained away, there remained our little garden by the door, still on its brick carpet, the water line just below the rim of the pots. So despite the dismay inside the house, we just stood about the door and looked shocked into the faces of flowers, watching the bees return. How people mentally survive being flooded in the winter I do not know; it must be one hell of a mind-freezing situation to be in.

So it pays to go big with pots for many reasons; little pots usually make a small space seem smaller and they require a constant nannying of watering and feeding because they get hot and restrain the potential of many plants rather than allow them to flourish – unless you want succulents, that is. For the record, I don't, not ever!

The only time I use small pots – old terracotta ones – is in the spring for small bulbs. They are also useful for planting up plug plants, and can look beautiful planted with violas, but even these will be far happier in larger pots as they are all about keeping their roots cool and moist.

The pots that I use more often, which are substantially bigger than small terracottas, are old coal buckets. These are much more generous, holding around 9 to 12 litres of compost. Even if all you have is a doorstep, plonk the biggest pot you can find into that space as it will make your gardening life so much easier and allow for a display that is as lavish as is possible. Most dahlias, annuals such as cosmos and sunflowers, and roses will only thrive in large pots; they need them to grow big root systems that will in turn result in masses of flowers.

Not all annuals are this demanding, though, and if all you have is a balcony then the amount of weight it can safely support will have to be considered. Window-box annuals that are happy to grow shallow root runs, such as nasturtiums, marigolds, cornflowers, linaria, panicum, violas and borage, will all thrive here (if they get some sun) and they will billow up and go over the confines of the window box.

BELOW Regardless of size, pots must all have drainage holes that are then covered with either broken pot shards, polystyrene or grit for drainage.

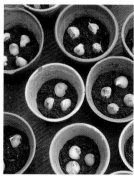

ABOVE Small pots are usually only used in the autumn for planting tiny spring bulbs such as iris, muscari and fritillary.

Herbs will be happy within small planters, too, but nothing will enjoy drying out to a crisp. The best lavender I have seen was growing in the top of a brick wall, planted in a narrow slip in the wall's top. It flourished because its snake of a home had a depth of rocky soil so the roots could grow down deeply into this chasm. Lavender is reliable and in the winter pots that hold it will form silver domes of the grey-blue foliage that will need to be cut back in the spring. It requires good drainage, but the soil can be a mix of sharp sand, grit and decent compost. Over the winter, goldfinches will enjoy eating the seedheads, unless you have collected them for lavender bags. To my horror, I keep seeing fake sprays of lavender in cafés and in window boxes lately – a huge sign of things going terribly wrong when herb royalty is replaced by plastic!

HANGING BASKETS

I'm not crazy for hanging baskets, but if you are going to go for these, absolutely go to town with them! Again, like pots, go for the largest basket that the bracket can hold – I like the ones that look as if they have been made from aged raffia and are of a deep triangular shape, as this allows more depth for the roots than ones that are bowl-shaped. Line them with pierced compost bags and several pairs of old socks before you add the compost, as these will soak up and hold moisture. *Thunbergia alata* looks especially nice tumbling out of a hanging basket and the fronds of *Panicum* 'Frosted Fountain' or 'Frosted Explosion' bubbling up and all over like a bird of paradise's crown. You'll see that I use panicum quite a lot for its Champagne froth-like sparkling seed heads. They remind me of the famous Champagne bar that I pass getting off the train when arriving into London's St Pancras station from Nottingham. I don't care for Champagne much though, gin is my poison if you're offering.

PAGE 16 The tulips here include the peony type 'Palmyra', the lily flowering tulip 'Sarah Raven', the classic orange and freesia scented 'Ballerina' and the plum smoke blush-toned 'Ronaldo'.

PAGES 18–19 Our home is attached to a derelict building whose back wall sheds white paintwork to give a glimmer of what would have once been a handsome red brick façade. I quite like its beaten-up look though.

LEFT *Thunbergia alata* 'African Sunset', a warmth-loving annual that really gets going in later summer when its bonnet-brown blobbed middles cheerfully open. It will climb through the foliage of dahlias providing good contrast to the dark foliage. Plant it out as a large, well-grown plant when the nights become properly warm; it will sulk otherwise.

All-important drainage

Regardless of whether they are large or small, all containers must have one thing in common: drainage holes. These are either made with a good clout nail or preferably drilled in, as this is gentler on the base. If pots cannot get rid of excess water the soil quickly becomes waterlogged and the plants within them rot – it is incredible how fast even a large pot will become a swamp if water is not able to drain away.

Before a pot is filled with soil, it is vital to cover the drainage holes with a layer of material that encourages the water to drain away swiftly and prevents the holes becoming clogged up. This layer of drainage, or clinker as we call it at home, needs to be a least a few centimetres (inches) thick, and made up of old crocks, broken roof slates, pea shingle or torn-up pieces of polystyrene. In a galvanised container, polystyrene is especially helpful as it stops the compost being in direct contact with the bottom and helps preserve it. If wet compost sits directly on top of it, it can cause constant rust, to which modern replicas of dolly tubs made of thin zinc are vulnerable.

Indeed, all plants, regardless of the season and their size, will be doomed to failure if their soil is waterlogged. I have seen dahlias in huge pots sulk and collapse within a fortnight due to their tubers rotting in constantly wet soil. The problem is that to begin with such plants give the impression that they need watering, so often they are given an extra drink. Always finger-test the soil first to be sure; if it feels wet, hold back the water and investigate, as the drainage holes of the pot may have become blocked.

But how much water should pots be given and how often? Climate change is making our springs hotter, so even this season's display often needs to be watered as if it were summer – once every two days if it becomes maddeningly hot. For the summer displays, watering lightly but often is important while the plants root into their pots, then, after two months, you should start to drench your pots with a full watering can. Never, though, water the foliage unless you are watering in the evening; water into the low side of the growing mass of plants. If you are using a hose then stand with the hose and count down from 50 to one, to ensure the pot gets a good soak. In the summer, saucers beneath small pots will allow them to soak up water from their bases overnight.

Once the summer show is growing well, the foliage will act like umbrellas and so any rain, even if it is heavy, often will not penetrate the soil surfaces, making watering at all times over the summer essential. If you have a downpipe on your house wall, consider giving room to an attractive urn-styled, fake terracotta water

BELOW Tin bath being prepared for autumn bulb planting. Drainage is essential for all bulbs and holes have been drilled into the bottom.

ABOVE One layer of tulip bulbs in a shallow tin bath, packed closely together, less than 2.5cm (1in) apart. Several different varieties are all mixed up together in a bucket and are then planted.

butt – one that you can take the lid off to then plunge the watering can into will make watering much quicker.

If you have a bath often rather than a shower in the summer, then using a long tube to syphon the water out of the window and down into the garden, just as if you were cleaning out an aquarium, is an idea to perhaps try in the name of sustainability!

LEFT Snow on the pots. The cardoons, kale and wallflowers take the elements in their stride. All the pots swiftly drain excess winter wet thanks to their drainage holes.

OVERLEAF

LEFT Tulip 'Brownie' is like delightful toffee popcorn. Here it is paired with an underplanting of Cleopatra-worthy viola 'Tiger Eye Red'. The buds of the late 'Black Parrot' will succeed this early flowering peony tulip.

RIGHT Viola parade in terracotta bowls and pots. Violas 'Honey Bee', 'Tiger Eye Red', 'Frizzle Sizzle Burgundy' and 'Frizzle Sizzle Yellow Blue Swirl' with faces like those of budgies.

Islands of flamboyance

If you have your heart set on terracotta you need to buy pots that have a frostproof guarantee, and also oxide within the clay, as this encourages a weathered look as the salts and calcium leach through. Terracotta is porous, so in winter, when temperatures fall below zero, they can crack. This can be a real annoyance if it is an expensive pot, but placing them very near to the walls of a house protects them, thanks to such microclimates being warmer. Raising them off the floor using the dinky terracotta pot feet that you can buy is good protection, too, as it ensures air circulation under the bases and improves drainage.

With small, newly bought terracotta pots, the best way to age them is to dunk them in a pond or in a water butt, then a coat of algae will grow quickly on them, Even a bucket can be filled up with rainwater for the task – it must be rainwater for the algae spores to be present. I have an upside-down dustbin lid propped up on bricks on the floor of the yard that we use as a bird bath and trios of terracottas take turns soaking in this, while providing little bathhouses to the precious town frogs! The water is emptied and refreshed weekly so that it does not harbour mosquitoes, and birds and bees also visit it to drink; all gardens should have watering holes for wildlife. After a month of being submerged, the teracotta pots will start to slime up and, once dried, will look marvellous.

Large terracottas too big to be submerged can be painted generously with organic, natural yogurt in the summer, which will then go green. For the algae to take, the pots need to be kept damp and away from full sun for a few weeks so that the spores can really get growing. When planting them up for summer, line their insides with old compost bags as this will help them to stay cool and reduce moisture loss.

DOLLY TUBS, GALVANISED BATHS AND METAL BINS

When I say pots, however, what I mean are galvanised baths, dolly tubs or old metal bins, not terracotta or expensive coppers – the latter would be stolen from a front garden sooner or later! I have come to realise, though, that not many people know what a dolly tub is. These barrel-shaped, ribbed-sided old dears were the original elbow grease washing machines used by almost every household in the nineteenth century. They would have been positioned by the back door and filled up daily with hot soapy water. To assist with the handwashing, they had a wooden dolly (a pole for stirring the clothes around in the tub) also known as a dolly-peg – in Yorkshire, the dolly tub is aptly called the Peggy tub! What I love about them is their shape, like

ABOVE Aging small terracotta pots by soaking them in a rain water-filled bird bath. Ours is an upturned metal dustbin lid. It is refreshed weekly to prevent mosquitoes but this small amount of water has attracted a frog, demonstrating water's power as an aid to wildlife, especially in urban situations.

OPPOSITE *Allium cristophii*, *Nectaroscordum siculum*, Martagon lily 'Sunny Morning', Oriental poppy 'Patty's Plum' and Eremurus 'Cleopatra' have enough scale to hold the garden in between the spring and summer show.

a hardboiled egg with its pointed top sliced off awaiting the dunk of a soldier. They accommodate everything that I want to grow in a garden well and comfortably. The downside is that they now carry a price tag that reflects their antiquity.

Much cheaper and just as helpfully big are galvanised dustbins. Try to buy them secondhand so that they are already roughed up and weathered down, as until they lose their shine they look awful. The old, thick metal ones that were once used as actual dustbins are the best, as they are made of proper galvanised metal, with good, thick handles on them.

Old cattle troughs are wonderful, if space allows, as they will fill whole corners of terraces and gardens like great beached whales. Seek out local farmers who will usually be willing to sell you old and leaking troughs that are already aged and bashed about. They are very heavy once filled, however.

Oval tin baths are lovely things, too, they look especially nice in rows sitting on a terrace or sitting on the edges of a raised bed. You will be able to find decent-sized ones in good condition for between 40 and 50 pounds; I hate making holes in them as I just imagine them being filled up and used as duck baths!

How much does it all cost? This is a question that I am often asked, particularly when it comes to the tulip bulb order, and it cannot be argued that gardening and spending money do not go largely hand in hand. Gardening becomes an addiction, albeit a good one, like cosmos and Champagne!

RIGHT 'Amazing Parrot' tulips in full flight with tulips 'Black Hero', 'Antraciet' and 'Black Parrot' giving striking contrast. The single pale 'La Belle Epoque' was a mistake but not an awful one. The scale of the cardoon leaves here echoes that of the tulips but they don't compete with them.

LEFT Crocus 'Spring Beauty' is very fairy-garden-like, with petals of a sapphire blue sharply diluted with the purest of proper purples. Several dozen in a pot will form a dazzling crown in mid-spring.

BRICK PATHS

The cost and daily work of the garden I justify as being necessary medication for the soul. The pots are the largest investment, but they came in dribs and drabs, in singletons and pairs; my mum already had a few dolly tubs, relics from old, now built-on allotment plots. Over the years, more have been added by me until avenues and clusters of them have formed, creating rows of contained flower beds. When I am looking for a new dolly tub, old bin or bath I will search all of the online market places I can think of as often these are the best places to find them at good prices.

Although expensive, plants, and spending money on them, is a cheaper affair than dealing with actual landscaping. Large pots can help avoid too much of this, but the herringbone brick of the yard is a saviour, a beautiful canvas, so hard garden surfaces are worth getting right. Reclaimed old terraced brick is classy, as are old slabs, but avoid the visual curse of new-build brick, gravel or wood chippings.

With reclaimed brick, do not trust the orange sort as these often have not been fired at a high-enough temperature. As a result, they have a tendency to become slippery and their faces crack with frost easily. Red brick is a better option but blue engineering bricks are by far the best for weather resistance; they are designed to be used underground, so they won't chip at all, but this will be reflected in their price; the same goes for any old bricks from abroad. The cheapest pathway of a garden could perhaps be hoggin, which is a mix of gravel, sand and clay that binds firmly together when compacted. If the right coloured sand is used it can look rather classy and natural. Vitally, it lets water drain through it and will also allow for some self-seeding of plants to occur. I imagine peahens pecking and dust-bathing in it within the garden conservatory restaurant of Petersham nurseries in Richmond, London. Here, it has been used underfoot in a Moroccan buff orange colour to glorious affect.

LEFT Fritillaries in small terracotta pots. These require watering every day, even in spring.

RIGHT Spring iris 'Purple Hill' and 'George' in an auricula stand. Their yellow flecks act as pollen guides which awakening bees can easily see as they take their first spring flights, looking for nectar and pollen.

OVERLEAF The garden in full tulip bloom. 'Prinses Irene', 'Parrot King', 'Flaming Parrot', 'Ballerina', 'Queen of Night', 'Black Parrot', 'Brown Sugar', 'Orange Princess' and 'Ronaldo' are complemented and made more luscious by the pot toppers of kales, wallflowers, violas, cardoons and primulas.

Compost for pots

Compost is the foundation of a container garden, so it is vital that you nurture and understand it. Compost needs to be alive and rich, full of worms, moisture, air and fibre. Think of it as needing to look like a freshly baked chocolate brownie and being like the topping of an apple crumble in texture. Drainage is vital to its condition also (see page 22). Unlike soil, compost contains a huge amount of humus (the organic substance that remains after plant and animal matter decays); more than would usually occur in top soil. Plants adore it and so compost mixes are loaded up with it.

THE PROBLEM WITH PEAT

Humus is found in wetland peat bogs. For decades, these have been plundered to make compost mixes, destroying what are crucial habitats and carbon dioxide stores. I once delighted in buying the big bags of cheap multi-purpose peat compost seen on sale in supermarket foyers, as they would fill up my ever-increasing array of large pots within moments, but no more. Aside from the environmental issues, the snag with peat-based composts is that their goodness is quickly exhausted. You need a more life-giving mix to ensure pot garden vitality. Avoid any claiming to be made of green waste as these often contain rubbish such as bark, which serves little purpose; they can also contain herbicide residue with damaging results. If you can afford it, organic composts are well worth buying to avoid any risk of contamination.

LEFT Refreshing and inspecting the soil in between replanting the garden at the end of spring and at the end of autumn is important to top up its nutrients and to check for signs of pests such as vine weevil.

RIGHT Arguably the best compost is homemade. A compost heap requires its rotting contents to be turned over with a pitch fork once a month over the summer to help aerate and speed up the rotting process. Claudia, a Blue Pekin bantam, rummaging for grubs amongst the compost heap.

HOMEMADE

I have no compost bin at home, mainly because traditional ones need to be sat on bare earth, which I don't have, also the plastic ones look awful and, in urban gardens, can attract rats. Up the road, though, lives my nana Min and here there is a bin in which I compost all the droppings from my hens, which I keep there too (see page 114). Hen manure makes good compost but it has to be given time to rot down as it is acidic when fresh. If I didn't have this extra space, I would seek out someone locally who sold compost from their allotment. I might even take on such a plot myself for the purpose of making my own because it really is black gold. It's what you want to get into your pots by the generous spade full each year not only to replenish nutrients but also to introduce earth worms to aerate the soil.

RECYCLING COMPOST

I do not throw away any compost; it is constantly enriched and remixed during the mass lifting and replanting. The replenishing occurs twice a year, when the spring bulbs are lifted and then again in autumm when the summer show is taken out. I have a large tin bath that is used for tipping the existing compost into so that it can be freshened up and aerated before being put back into the pots. I end up with a few buckets spare each time but I save this and use it for potting up seedlings. For seed sowing, you should use a specific seed sowing compost as it will have been sterilised.

MAKING PEAT-FREE COMPOST WORK

I have trialled several, including those made from coir and sheep wool. The results have been mixed, particularly with dahlias. Some encourage an abundance of foliage at the expense of flowers. My view is that we should support them and that in time, with more industry effort, they will get better, but until then it is best to add homemade compost and soil conditioners to bulk up the levels of nitrogen, phosphate and potash. Generous handfuls of organic chicken manure pellets, Grochar fertiliser and comfrey pellets will all help to ensure a lush display.

Feeding the summer show with a liquid seaweed each week while watering will ensure a constant top up of nutrients too, resulting in vibrant foliage as well as abundant flowers, especially from dahlias and cosmos. Worm casts are also helpful for breathing life into stale and spent compost. I would avoid adding horse manure as it often contains weed seedlings unless you buy it sterilised. Homemade compost needs to be checked for white weed roots. These are likely to be of ground elder and bindweed. I once collected the tempting black soil that results from lawn molehills; a few trowels of this will be beneficial but can contain weed seeds.

Peat free composts are helpful in the summer as they hold onto moisture well but this can be troublesome for the spring display as bulbs prefer light compost mixes, so you will need to add horticultural or poultry grit.

ABOVE Earthworms in pots signal that the soil's chemistry is good. Worms will appreciate annual applications of organic compost. Leaf mould is rich in humus and easy to make yourself. Stuff leaves lightly into jute sacks in the autumn. Keep these moist, in a corner and the leaves will rot down within a year. Unlike a black compost bin, these sacks will not look unsightly. Ash, beech, hornbeam and lime leaves all make good leaf mould. As these are popular street trees, leaf piles in the autumn will not be hard to find – just ignore the funny looks you may get while you gather them up!

PERSIAN CARPETS

Exuberant, luscious and elegantly wild, these are the flower personalities and colours I grow.

My garden's orgy of a floral rainbow is made up of rich, egg yolk orange, which is the best smile-inducing colour, along with blush apricot and an emperor's robe, lemon curd yellow. Reds are varied, going from cardinal, pillar box red to rich sunset, glowing ember corals and ruby clarets, on into mahogany racehorse stud bourbon and fox fur rust. An occasional sprinkle of silk fuchsia and sugar plum fairy purple is to be found too, as well as indigo and peacock blue.

It is important to form a palette of your own favourite colours that will make your heart sing and your mind buzz. When putting mine together, I think of sun-ripened fruit bowl tones and of rich, antiqued Persian carpet tapestries. What I avoid are factory sweet colours as these are too brash and chemical, and white is totally banned, as are any milkshake pastels. These would ruin the carnival, samba dance bravado.

This diversity of colour and shape forms an erupting coral reef. A bizarre yet gorgeous collective of Vivienne Westwood-like dresses, an outrageous Muppet rabble, zinging off one another's presence from pot to pot. There are crazed stars, burning flames and sparkling fascinators. These floral beings have charisma and bravado, they're other worldly, offering a visual ecstasy and an escape into Wonderland.

Planning

I like gardening with annuals and bulbs because the pots can primarily be treated as if you are curating two different displays every year, one for spring and the other for summer. Essentially, the garden's key show is one of tulips, then dahlias. This is my garden's calendar of flowers through the year:

EARLY SPRING – crocus and iris with the foliage of kales, fennel and cardoons.

MID-SPRING – tulips en masse with the above foliage along with wallflowers and kale. These plug the late spring gap being large, individual flowers.

EARLY SUMMER – The tulip bulbs are all removed delicately, leaving the alliums that will be in bloom by now in the same pots that they were in to plug the late spring gap. The alliums are supported by poppies, lilies, and the towering rockets of eremurus. The leaves of the cardoon seedlings give a large presence now, too.

MIDSUMMER – The sweet peas having been planted in separate pots back in spring flower en masse. Meanwhile the other pots are replanted completely with the summer show of dahlias and annuals.

SUMMER TO LATE AUTUMN – dahlias, cosmos, sunflowers, quick-flowering half-hardy annuals and annual grasses provide a jungle that flowers until the first frosts of winter.

WINTER – the foliage of kales, wallflowers, cardoons and fennel along with birch woven supports give form to the sleeping winter garden, with most pots having been planted with hundreds of bulbs for spring.

It's a constant space of planning and regrowth and although labour intensive it is massively nurturing to creativity. I always try a new variety of tulip, dahlia or annual and look forward to seeing if I like it. If I fall in love with it, fantastic, if I don't, then not to worry as I can just take it out. After a few seasons of growing a garden like this you will create a sort of personal Noah's Ark of seeds and bulbs that you really love and trust to create the feel and look that you want, the packets of seeds and bags of bulbs that you would take to a desert island.

This is why I don't have a garden full of pots that also hold box balls or lavender, because if I had these too I wouldn't be able to grow a garden with peak shows of spring bulbs and summer annuals and dahlias. I like every pot to be of the same consistency of growth, otherwise the garden wouldn't be at the same level of abundance. The downside is that the garden is seemingly lifeless in the winter, but I don't mind this, knowing the promise of floral eruption that each pot has in it like a ticking clock that will chime as soon as the horizon of spring is felt.

PAGES 36–7: Dahlia 'Schippers Bronze' with black elder-like foliage, and dahlia 'Totally Tangerine', paired with the annual grass Panicum 'Sparkling Fountain'.

PAGE 38 Tulip 'Amazing Parrot' is a deep coral pink merging with orange. Cardoon foliage provides an equally commanding leaf escort.

OPPOSITE Clearing the bulbs with help of a Barbu d'Uccle Millefleur bantam hen.

OVERLEAF The garden is planned using mood boards for spring and summer. Each pot is drawn on the plan and then all the chosen varieties of annuals to be grown from seed or bulbs are laid out to see if they will look good together. An immensely helpful process both creatively and practically.

Spring 2019

Boxing Day
Sown Sweet
Peas —
Natucana
Blue Velvet
Windsor
King Ed VII

a

None

'Ballerina'

Tulip 'Queen of Night'

SWEET PEAS
SILVER BIRCH.   x 2 DOLLY TUBS

'Sarah Raven'

C. 'Spring Beauty'

Iris 'George'

'Windsor'   'King Edward VII'   'Matucana'   'Blue Velvet'

'Black Parrot'

Crocus
Hyacinth Woodstock
Fritillary

'Prinses Irene'

Cartoon
Seedlings.
→ All

Wallflower 'Sunset Bronze Shades' F1

Blood Red Covent Garden

Kale red bore
Sunset wallflowers
Bronze shades
All

T. Ballerina
T. Queen of Night
T. Black parrot
T. Ronaldo
T. Princes Irene

'Fireking'

T. Black Parrot
T. Black Hero
T. Amazing Parrot
T. Antraciet
T. la belle Epoque
T. Sarah Raven
T. Victoria Secret.

Cartoons + Bronze
Fennel

PATH

I. 'George'

— Alliums
  — Christophii
  — P. Sensation
  — Nectaroscordum

'Ballerina'

'Sunset Red'   3.

'Ronaldo'

A. schubertii

DOLLY TUBS

Front Door

Dianthus barbatus 'Sooty'

'Palmyra'

Dahlia
'Black Jack'
Red millet
glatioli
'Espresso'

D. Waltzing Matilda'
Glatioli 'Espresso'
D. 'Schippers Bronze'
Panicum 'Frosted Firework'
Plectranthus'

'Schippers Bronze'

OVAL ZINC BATH

D. Waltzing Matilda'
Dahlia 'Bishop of 'Aucklan
Panicum 'Frosted Firework'
Plectranthus

Dahlia 'Black Jack'
Cosmos 'Rubenza'
Red millet

Totally Tangerine

Cosmos 'Rubenza'

D. Soulman'
D. Totally Tangerine'
Panicum 'Frosted Firework'

Chocolate cosmos
D. 'Dark Desire'
Pletranthus

RAISED BED SIDE

DOLLY TUBS

'Plum Tart'

Dahlia
Bishop of
Auckland'   Replacing
Red millet   sweetpeas
Pumpkin   in July
'Black Futsu'
Glatioli Espresso'

D. Soulman'
D. Totally Tangerine'
Panicum 'Frosted Firework'

PATH

Cosmos 'Rubenza' Sunflower 'Claret, red millet
glatioli Plum tart

D. Waltzing Matilda'
Chocolate cosmos
Panicum 'Frosted Firework'

Dahlia 'Black Jack
Red millet
Glatioli 'Espresso'

'Happy Single Kiss'

FLOWER BED

Soulman

Dahlia
'Bishop of Auckland'
Red millet
glatioli 'Espresso'
Cosmos 'Rubenza'

FRONT DOOR

DOLLY TUBS

CUTTING UP PLANT CATALOGUES

Cutting up plant catalogues pushes your plant creativity and helps you realise what colours you really like. Some catalogues are total misleading trash, recycled straight away, but others are worthy of being read, having been produced like mini books. The highlighter is to hand, photos and names are underlined, and a list drawn up – often a very big one. This you need to streamline, looking at when each selection is going to be in flower and what will take over from what. Then you should research further, typing the varieties into Google and Instagram to be sure that you really like them, and that the selection, especially those that flower in late summer and early spring, is truly attractive.

I then plot the garden with circles representing all the pots. The catalogues are cut up and the faces of the chosen flowers laid out. Which ones will be the season's main players? Three to five will be grown en masse to give the garden a constant flow of similarity. Which will be the faces that burst out from these main players like regal figures? And which plants will give the essential leaves, seedheads and tendrils to form all the flowers together within the pots? All these questions can be laid out and glued down until the entire show has been selected.

BELOW Instagram posing with the dahlia 'Emory Paul', which is like a flamingo wonderland croquet mallet, gorgeous in some ways but its flowers do look almost painfully ridiculous with some reaching the size of useless footballs upon what are quite tall stems. This really is too much, or it would be if the red millet sprays weren't hiding my pout.

Embracing colour

Floral pattern is useful inspiration to ensure visual splendour in a garden. You need repetition of similar colours and shapes, with the occasional flare of something slightly bolder and different but not something that commands attention stupidly away from everything else – combinations of flowers that are complementary but still exciting, with pop and zing but also flow.

A pair of growing 'curtains' being drawn day after day is apparent in the summer when this season's growth reaches towering heights and jellies out over the brims of the pots. The foliage and flowers merge toward one another so that the path to the door is almost hidden, gloriously so, to the point that you'd be forgiven for thinking the garden was just grown in flower beds.

This is why it is important to maximise your space with the largest containers you can accommodate; small pots cannot give the root space plants need to grow to their full, door-engulfing potential. While curtains of William Morris swirling dahlias are possible in summer, the same level of height and growth cannot be achieved with as much depth in early spring. This season starts off as a lower, one-level Persian carpet affair. It relies on the faces of flowers and leaves as individual characters rather than as a deeper, collective flower and foliage show that the speed of summer growth can quickly nurture. But then out come the first tulips and suddenly the garden's mood goes up a gear into high pizazz and craze.

This tapestry is not uniform but it still contains repetition, not in lines or lumps, but by encouraging naturalistic flocks and swirls. The tallest dahlia is not planted in the middle of a large dolly tub, but on the edge, and nothing is made to look unnatural. Layering height within a pot is essential, canopies of colour and form that merge together and give succession through the spring and summer seasons, and also help to hold the garden together between the end of spring and early summer, a gap when the spring bulbs are largely lifted and the summer display is planted out.

Avoid thinking you can have one pot in stained-glass tones and another next to it in ivory white, because in a small space this doesn't work. You cannot divide a small garden into different rooms to give the separate colours the space they need to breathe or gently merge together. It is better to select and be loyal to colours that go together, such as reds, purples, oranges.

I get irritated by pastels. I avoid anything that reminds me of deathly dull places, namely waiting rooms at the doctor and dentist, care-home shades or, worse, stained mattress shades – which seem to be all the rage at the moment!

OVERLEAF Looking down on the foxtail lilies, *Eremurus isabellinus* that flower in early summer. 'Cleopatra' is the best variety of apricot and tangerine orange. Their spires, which are made up of thousands of star-shaped flowers, open from the bottom upwards. Amazingly, these are strong against the wind and rarely need to be staked. Thanks to huge amounts of chicken grit, to provide essential drainage, they can thrive in pots. The only drawback is that, like alliums, their foliage becomes tatty by the time that their bee-intoxicating floral towers come into full bloom. They will easily sulk, too, resulting in no flowers, but the presence they give, if they are content, to an early summer garden makes up for this. See page 176 for planting.

OPPOSITE There is a colour rule of no white in the garden, except for this pot, which flowers before all the others start to awaken in late winter. The dwarf cherry, Prunus incisa 'Kojo-no-mai', is slow growing. It is underplanted with the perennial appearance of crocus 'Jeanne d'Arc'. The pot's top is dressed for spring with raked up lawn moss.

LEFT Floating snowdrops 'Flore Pleno'. The best way of admiring these flowers is to float them in water. They will last several days like this.

WHITE CAN BE A COLOUR

White on its own looks resplendent but it has to be alone in terms of the flowers and complementary foliage of greys, eau de nil greens and silvers. I will admit that this all sounds incredibly rigid and that sometimes it pays to break such rules. I recently saw a video of the bearded irises in Monet's garden, all of different colours like hundreds and thousands, and it was, admittedly, beautiful, so sometimes it pays to throw colour caution to the wind, but not if you have a small garden!

In early spring, I allow white to flower in the garden because it has the space to itself before other colours begin. This white pot is one of the few that is rarely disturbed or replanted. At this time of year in the cold, dusky light, white looks ethereal, untouched, graceful and ghost-like. Make no mistake about it, if you want purity go for a white garden, especially if you are a night owl, when white really becomes otherworldly en masse, just as wild cow parsley looks on a country lane in the twilight.

This white pot contains a dwarf Fuji cherry that has a compact and crazed over-grown canopy of tiny branches, all twisting inward, gnarled and magical, and before any leaves appear comes the purest of tiny blossom, a wedding cake-like creature that honey bees delight in finding. At the base of the pot, to complement this, are huge crocuses, fabulously christened 'Jeanne d'Arc', but before these and the cherry blossom, the pot holds court to the large and double snowdrop blooms of 'Flore Pleno'. These swing from their thin stems like lanterns on pendulums. It is worth cutting this snowdrop and floating the stems in a little goblet of water so that you can admire their beautiful, otherwise hidden, faces early in the year.

Floral curtains

Once, I was feeling rather giddy and found myself in the fabric department of John Lewis because my bedroom window needed curtains. It still does to this day. My jumbled-up, crazed shell of a room that is often a greenhouse needed, I felt, some attention. I began looking through the rails of fabric samples, somewhat carelessly. I was quickly entranced by one featuring moulting peonies, fully blown roses and flirting parrot tulips swirling about lavishly on a midnight black background, looking like seaweed in a murky, velvet rockpool.

The fabric, of Osborne & Little pedigree, was chosen; giving the measurements of my tiny bedroom window, I asked for a quote.

ABOVE Kale 'Redbor' and tulips 'Ballerina', 'Black Hero' and 'Black Parrot'.

OPPOSITE Dahlia foliage 'Black Jack', 'Schipper's Bronze' and 'Bishop of Auckland'. Gladioli 'Espresso' and 'Magma'.

Obviously, I wanted them to generously puddle! My hair was bouffed up as much as my hands could manage into my impression of *Absolutely Fabulous*'s Patsy Stone. The lifestyle of Patsy echoes my attitude to interiors. She adored her soulmate friend Eddie's house, coping without food quite well, parasitic really! In the end they found peace, living in The Conran Shop window; before this she lived in a grey shell, above Oddbins!

With these *Ab Fab* clips going through my head, the store assistant tapped all my curtain requests into her computer and we waited for the price result to be calculated. The quote (thankfully it was just a quote) was of a figure large enough |to buy several dozen pot-grown David Austin roses with enough spare change for organic manure, or it could have paid for several years' worth of tulip bulb orders.

With my hair still in beehived, albeit taken-aback contemplation, I asked for a fabric sample to please be sent, in a 'cheers, thanks a lot' tone.

A pitiful patch of fabric duly arrived weeks later. A little square that now lines the bottom of an empty wooden bird cage in memory of the strange time that I took an interest in interiors. So my room still is without proper curtains.

PREVIOUS PAGES The same pots but showing their spring tulip and summer dahlia displays.

LEFT Cardoon leaves and tulips 'Ballerina', 'Ronaldo', 'Queen of Night' and 'Black Parrot', the last two of which are in bud. At the front is the early flowering 'Palmyra'.

RIGHT A trifle-like ooze of summer colour. Dahlias 'Waltzing Mathilda' and 'Bishop of Auckland' with *Thunbergia alata* 'African Sunset' creeping through them and chocolate cosmos with gladioli 'Espresso'.

LEFT Dark-leaved 'Schipper's Bronze' dahlias are especially helpful in providing good foliage.

RIGHT A party of snapdragons in rich, childhood rocket-lolly colours. These are a mixture of a series known as the Sonnets. They are good both as hard-flowering pot plants and for cutting. Starting to cascade under them are calibrachoa.

TULIP CATWALK
TOP ROW LEFT
TO RIGHT
The late tulip 'Black Parrot' looks like a reptilian egg in bud before hatching into a purple bird of paradise. Tulip 'Queen of Night'. Tulip 'Orange Princess', a peony tulip with petals of the truest vermilion orange. Tulip 'Cairo', a single and early tulip. Tulip 'Black Hero', a tall, late tulip that ruffles out despite starting out as a small, conker-like bud.

MIDDLE ROW
LEFT TO RIGHT
Tulip 'Irene Parrot', a short tulip, bred from the classic 'Princes Irene'. Tulip 'Ronaldo', an early tulip, its petals have a blushed plum smoke-like finish. Wallflowers 'Sunset Orange' and the scarlet 'Vulcan' below provide scent.

BOTTOM ROW
LEFT TO RIGHT
Tulip 'Queensday', whose flowers are made up of blood orange coloured petals. Tulip 'Bastia', almost ugly, but intriguing due to its crystal-like fringed petals. Tulip 'Palmyra', the earliest of the peony tulips and of truest, satin mulberry red, like the ballgown worn by Keira Knightley when she played Anna Karenina.

Sweet peas

The quintessential cottage-garden annual flower. Romantic, strongly scented and the icon of the home-grown cut-flower movement. No flower garden should be without them. The growing of them is one of the most passionately debated things as everyone has a better way to do it, but this is my method of growing in them pots.

My sweet peas, like everything in the garden, grow in pots, in the biggest and tallest of the metal dustbins. Growing them in something large is essential for them to match the towering flower splendour that is seen when they are traditionally grown in the ground. For this reason, I don't plant bulbs in two or three of the bins, because by mid-spring the sweet peas will want to be planted out.

Sweet peas are reliant on having a strong and long root run, as it is these roots that will then fuel the plants to flower well. The roots of sweet peas always need to be constantly unrestricted and be able to access nutrients; if you can ensure this then your sweet peas will flourish. These important powerhouse roots need to be given time to grow; for this reason, the seeds of sweet peas should be sown in the winter not in the spring because this cold weather stalls their upper growth and puts the seedlings' focus into growing their roots. Some people sow them earlier in the autumn but I have found that doing this at the beginning of winter works well in the windowsill's seed timetable, as there is nothing else on it at this time of the year.

There is such an array of sweet pea seeds it can be overwhelming to choose. Like tulips, they come in a vast colour range so there are varieties to suit any garden's colour scheme. It's their growth habit of a naughty tendril, almost whip-like floral bower that make them irresistibly romantic and fun to grow every year.

RIGHT Sweet peas flowering at full pelt require daily cutting, dead-heading and feeding to keep them looking good. In late summer they tire and need to be cut down. They can be replaced with late sown end of spring sunflowers.

OVERLEAF

LEFT A wigwam of stained-glass-coloured sweet peas – 'Blue Velvet', 'King Edward VII' , 'Matucana' and 'Black Knight'.

RIGHT A year of growing several marbled types. 'Wiltshire Ripple' and 'Earl Grey' are recently bred with a good length of stem and are deeply scented too.

LEFT 'Black Knight', an old variety of sweet pea of the best deep claret colouration.

RIGHT Sweet pea 'Almost Black'.

LOOKING AFTER SWEET PEAS

Each dolly tub takes seven sweet pea seedlings; this sounds like a lot but it ensures a good, well-covered wigwam, and with lots of feeding and watering the plants don't mind being close to one another. The tendrils at first do an important job of attaching their growing stems to the wigwam, but as the first flowers open the stems that are growing on the outside of the wigwam will want to be tied in with hessian string, otherwise these will often bow down and snap.

As soon as the first flowers start to bloom, picking has to begin. The tendrils should be picked off too, as these by now have done their job and those that have not attached themselves to the wigwams will begin to smoother emerging flower buds. They will take energy away from flower production, so remove every tendril and flower that you see once a week. A handful of flowers and a handful of tendrils is a good thing to come away from a sweet pea wigwam with.

Feeding the sweet peas is vital, well-fed and well-watered will see the sweet pea wigwams look good until late summer. The best sweet pea feed is made from comfrey, which can be easily done at home but it does smell so I do a bucket of it in furthest corner of the garden away from the door!

Comfrey, *Symphytum officinale*, can often be picked at the side of roads, on waste land or someone you know may have a clump of it on their allotment. Pick a good amount of the hairy leaves – a carrier bag's worth – and cut these up with scissors into a bucket when you get home. Fill the bucket with water and allow the leaves to ferment for three weeks. This will result in a truly foul and gag-resulting brown broth that is best covered over with a dustbin lid for its entire existence; it will attract mosquitoes otherwise.

Although it seems like a witch's brew to us, this tea is nature's best form of potash and it is completely free. A mug's worth added to a filled watering can to dilute it can then be fed to everything in the summer garden. It is a gin and tonic of goodness, especially to sweet peas, and will help them to fight off mildew until they eventually tire out.

When the sweet peas do go over, cut the stems off the canes and remove the wigwam – you can lift it out from the dustbin and strip it down, or it can be reused to grow pumpkins up. Do not dig out the sweet pea roots, just fork the bin over and add some fresh compost; the roots will rot into the soil and aid its structure. See page 168 for sowing sweet peas.

ABOVE Sweet pea 'Matucana' is the strongest of the old sweet peas in terms of its scent but alas the flowers are held on very short stems.

RIGHT Loose tendrils take energy away from plants and they grow tough and large, serving no purpose. They should be pinched off every time flowers are cut.

Dahlias

Dahlias encompass razzmatazz and cut-flower harvest productivity like no other plant. The more they are cut for the vase, the more flower buds they produce, and no other flower gives as much glamorous clout en masse. They have deservedly become the most popular summer and autumn cut flower, enjoying a revival over the past few years that has not been seen since they first crazed Georgian England as new garden introductions from Mexico.

They then fell out of fashion, with the oldest varieties available today being kept alive thanks only to a few dedicated growers who continued to show them before this huge garden resurgence took hold.

Last year, I stood in the middle of a dahlia bed planted just with decorative types, the huge-headed mass-petalled ones that most people think of when they imagine a dahlia. While some were indeed beautiful, I was horrified at the lack of bees and butterflies; the place was totally silent as it was just full of manmade, sterile flowers. I prefer, in the main, to select dahlias that are not these classically decorative types, but that are bee-attractive.

A single dahlia's heart of stamens and anthers quite literally looks like a honeycomb on a hot day and so, unsurprisingly, bees and butterflies flock to them en masse. The garden really does have an almost constant haze of pollinator activity in late summer and autumn when few other nectar-rich flowers are to be found. To see such a feast is soul-nurturing. Dahlias that are of the right varieties really can provide an essential lifeline at a time when many pollinating insects desperately need to fuel up before they either migrate or hibernate.

The single and anemone varieties provide this, and they are not just bee-worthy but garden-worthy too in their performances. They flower prolifically, often coming into flower weeks before their larger-headed, profusely petalled cousins. Most of these dahlias also have attractive foliage of dark green and bronze tones, far more voluptuous for a pot than just a dahlia with the body resembling that of a green cabbage! The dahlia 'Schipper's Bronze' looks like a crimson elder, while the classic 'Bishop of Llandaff', of cardinal red, and 'Bishop of Auckland', of deepest claret, have leaves and stems that are the colour of an After Eight-mint box, an almost oil black of green mallard drake-head shimmer.

Earwigs are less of a problem with the single dahlias (they do not have the profusion of petals to hide in) and anyway will help to eat any aphids that gather on the flowering stems. If earwigs do prove to be an issue, the old-fashioned method of

LEFT Dahlia 'Black Jack', one of the most fabulous of the decorative dahlias to have in a large pot, thanks not just to its huge flowers but to its dark leaves and stems too. Each stem will require staking due to the weight of the flowers.

a cane and upturned terracotta pot stuck to it using Blu Tack (or the equivalent) and stuffed with straw will see them gather in it each morning. They can then be emptied out somewhere, I don't quite know where to suggest that you do this, especially if you just have a balcony garden, but find somewhere away from the dahlias!

I do grow a select few of the large, decorative, cutting dahlias though, as I cannot resist having these big party-like hats in the garden, and they have a much better vase life than the singles because of all their extra petals.

'Black Jack' is my favourite dahlia for both cutting and giving the garden a big statement piece of a flower. It is quite newly bred with the glossiest, varnished maroon flowers. These are held by dark stems with deep green kelp-coloured leaves. So far, the foliage qualities of the decorative dahlias have not been given much attention. I hope that the introduction of this variety will encourage more to be bred with attractive foliage.

The classic Sarah Raven stalwart dahlia is 'Rip City'. Sarah, the queen of dahlias and one of the first gardeners to really grow dahlias both as cut and garden flowers, discovered it in Monet's garden in Giverny; its flowers still are yet to be beaten in looks. They are like thick, claret, chocolate icing whipped on top of a cupcake crossed with the shape of a frenzied sea urchin, with its curling tendrils trying to filter out morsels in a frothing tide – absolutely delicious temptation and begging to be cut as a single-piece flower crown. If you have thick hair it will look amazing shoved in and clamped somehow, using a clothes peg, trust me!

Then there are the tight-ball dahlias, the grandly titled 'Downham Royal' of deep racehorse claret and the purple 'Jowey Mirella'. For mandarin orange, 'New Baby' is the one to grow of this class. These have a superior vase life compared to the day event that the singles often give when they are picked. No dahlias, however, will last more than five days in the vase, and, as a rule, they do not continue to open once cut either, so they must be picked when they look good in the garden.

Nonetheless, all dahlias thrive on having their flowers cut. Pick them weekly and they will produce more and more. If you do not cut them you will have to dead-head them, which is a thankless task in comparison. For flower productivity and earliness to flower, 'Totally Tangerine' and 'Waltzing Mathilda' take some beating.

The single and anemone dahlias tend to be small to medium-sized in height, growing to around 75cm to 1m (30 to 40in) tall. This height makes them perfect for a dolly tub, which will take three dahlia plants, or a large window box could also hold them. The other bonus of growing this group is that they don't grow stupidly huge tubers (the hard, sweet-potato-like callus roots from which dahlia stems sprout), so they are easily lifted each autumn and potted up again each spring. Some of the decorative dahlias, in comparison, grow huge tubers within a season, so are then almost impossible to remove from a dolly tub or bin, requiring a spade rather than a trowel to do so. See page 172 for tuber planting advice.

ABOVE 'Downham Royal' ball dahlias, while being useless for bees, are very good as cut flowers, as they last and travel well.

LEFT Dahlia 'Rip City' with *Panicum* 'Frosted Explosion'. Planted in huge terracotta pots framing one of the potting sheds in the garden at Chatsworth House in Derbyshire they prove that, even in a large garden, pots give spaces floral exclamation marks. Chatsworth is one of my favourite places to visit, especially the kitchen and cutting gardens that are looked after by Glenn Facer, Becky Crowley and Sophie Bromley.

OVERLEAF

LEFT 'Soulman', an old variety of anemone dahlia with dark mulberry-jam middles radiating out to velvet magenta petals, suitable for a large pot with staking.

RIGHT A jumble of favourites. 'Bishop of Llandaff', 'Waltzing Mathilda', 'Rip City', 'Hillcrest Royal', 'Bishop of Auckland' and a French marigold.

PERSIAN CARPETS

DANCE OF THE DAHLIAS
TOP ROW LEFT TO RIGHT D. 'Blue Bayou' is especially beloved of butterflies. D. 'Happy Halloween', a deep orange ball dahlia. D. 'Mango Madness', a zesty pale yet zingy yellow dahlia. D. 'Sarah Raven', a striking anemone dahlia. Cosmos 'Rubenza', whose flowers open as a lavish scarlet. Pick and deadhead weekly to ensure vitality.

MIDDLE ROW LEFT TO RIGHT D. 'Bishop of Llandaff', with flowers of striking post box red. D. 'Bishop's Child' – a packet of these seeds provides a rich, bee attractive mix of orange, purple, yellow and scarlet flowers. D. 'Schipper's Bronze' growing with the claret D. 'Bishop of Auckland' above it. D. 'Verrone's Obsidian', a velvet shooting star.

BOTTOM ROW LEFT TO RIGHT D. 'Totally Tangerine' – hugely prolific flowers of peach petals with a rich orange middle. Petunia 'Red Valour' will provide satin rich clouds. D. 'Bishop of Canterbury', the best raspberry jam deep pink contrasted by dark foliage.

Dahlia companions

While dahlias will hold their own in pots, there are many annuals that can be good supporting artists to the single and anemone types within them too. Try to plant these within a few weeks of the dahlias being planted. They need to be growing well in 9cm (3½in) pots so that they don't just get swamped when added around the sides of a pot.

Annual grasses such as panicum and millets will provide vertical spears and their seedheads will marry with the full-pelt flowering of the dahlias in the autumn, creating a pot that is a feast for bees and birds alike.

Cosmos 'Rubenza', the richest of the carmine claret cosmos with an airy habit, can either be grown in a pot to itself within an avenue of dahlia pots, or can be planted with the dahlias threaded through each pot as if they have self-sown themselves.

Thunbergia alata 'African Sunset' will, if planted out as a substantial-sized plants, snake through the dahlia foliage, seemingly disappearing for a few weeks, then they will delightfully surprise as their bonnet-like faces open in a mix of pink and orange-blushed apricot shades that look especially good against the dark leaves and stems.

Sunflowers Annual Helianthus are fast growing and complement dahlias well. Recent breeding has seen smaller varieties emerge such as 'Ms Mars' which grows several, smaller sized flower heads rather than a traditional single, dinner plate sized one. This has made them more garden than just allotment, tallest sunflower competition worthy. Traditional varieties such as 'Claret' are best planted in their own large pots. Sow them in early summer to replace the sweet peas in midsummer.

LEFT A honey bee on a chocolate cosmos. Like dahlias, these too are tender tubers. Their flowers dance and sway, held delicately on tall stems; they are happiest if they are given their own pot where they will not be overshadowed by any clumpier annual neighbours.

RIGHT A stained-glass window of jewel colours in high summer – 'Waltzing Mathilda', 'Bishop of Auckland', 'Blue Bayou', 'Soulman', 'Totally Tangerine' and 'Happy Halloween'.

OVERLEAF Late sown early summer sunflowers 'Procut Plum', 'Ruby Eclipse' and 'Claret'. Flowering in early autumn, they have tall stems of silver birch pushed in all around them to act as natural, loose supports from the wind. Their seed heads will become natural bird feeders. Behind them an arch of Cobaea scandens is starting to flower.

Pumpkins

Growing pumpkins is purely a visual affair, to give the garden just a little squeeze of Felicity Kendal good life. This is shameful, first-world vanity vegetable growing for still-life personalities and garden presence, but it is not at all part of the obsessive, size matters, polished and weighed ambition associated with growing show-bench-worthy individuals. I do not call them squash ever because all the fun and magic that come with the title of pumpkin is lost on this word. Squash is a drink or some sort of ball game. Just like sunflowers, pumpkins embody the energy and growth of life as they clamber and their fruits swell in such a beautiful array of yellows, greens, reds, oranges – even blues – and now in wrinkled peach, too!

Crazed, almost ugly fruiting-cucurbit creatures are welcomed here, the sort of pumpkin freak shows that I am sure Helena Bonham Carter, who we saw divinely in masses of silver chiffon as Cinderella's fairy godmother would, I hope, delight in. I am rarely growing for taste; I want fairy tale fruits of character, almost vegetable pet-like name-giving qualities – how could you consider taking a knife to them? I am on the verge of putting a pumpkin in a pram, do arrest me officer; this is my sort of concept. When my baby brother Lyndon came into the world, it was as if he was stuffed full of Crown Prince pulp; he was the perfect sort of fat cherub Cabbage Patch Baby!

I like a very subtle mix of select vegetables growing with flowers. It is very exciting seeing a fruit or pod growing by the day. They are different creatures to a quicker-opening flower bud; there is a more gradual garden performance of leaf than budding, which leads to flowering before fruiting and then ripening. Each

LEFT Pumpkin 'Black Futsu' is a gorgeously unusual pumpkin of blushed apricot. This variety takes time to get into its growing stride.

RIGHT The 'Crown Prince' pumpkin is growing up a hazel grid with thick branches in a noughts and crosses fashion. 'Earthwalker' sunflowers are also growing in the same tin bath. Dahlias 'Totally Tangerine' are in a pot in the foreground.

stage is captivating. Some pumpkins, aside from the giant orange sort, are visually useful; the varieties I like are those that are climbers with tendrils that pull their stems up, and in a small space choosing this sort of breed is essential as you cannot afford to grow those that prefer to take over ground space. Elevated sorts, however, will use free vertical space, taking over wigwams from fading sweet peas in mid-summer, which is when you will notice their growth goes up a gear because they get going as warmth grips the nights properly. You can use them to cover shed roofs and porches and they will do this willingly, provided their roots are not starved ever of water or feed.

Their large flat seeds need to be sown in spring on the windowsill; a 9cm (3½in) pot for each one will form their first nursery. It is pointless sowing them any earlier as they truly need the warmth of the oncoming summer and they cannot be put outside until it is properly warm at night.

You need only to grow a pair or a trio of pumpkin seedlings for a small garden. They are very hungry plants and each baby needs as much space, light, water and feed as you can give them. This will ensure that they not only fruit well but look good too. If they struggle or are deprived of nutrients their leaves will quickly develop powdery mildew and the fruits may fall off before they ripen, but with nurturing they will give the garden big, jungle-like leaves and tendrils.

Their large, very lovely yellow flowers are uniquely scrambled-egg-like in form, and in being both male and female. Bees are essential, of course, if the females are to produce fruit. You will see the baby fruit behind the flower in the female, if successfully pollinated her flower will fade to leave the fruit to swell and ripen.

For the seeds to awaken at all, they need to feel cosy in their compost, so they will sometimes sprout faster if they are moved to somewhere warmer inside, such as the airing cupboard or on top of the boiler. As soon as the seeds sprout, put them back onto the windowsill.

Two big Dumbo-like ear leaves will open out first. Be careful not to water the seedling too much until it forms its first true pair of leaves, as overwatering can cause young plants to rot. The first true leaves will look different from the first ones as they will have little hairs and crinkled edges rather than being rounded and smooth.

Pumpkin seedlings will want to be potted up into a 1-litre (2-pint) pot after a fortnight of being on the windowsill. Start putting them outdoors in the day and gradually harden them off.

Once roots are showing through the holes in the base, the young plant should be planted in its final pot, where it can grow into adulthood. A whole dolly tub or bin now spare from gone-over sweet peas is perfect, with the stripped sweet pea wigwam put back into it to be reclimbed by the pumpkin. For a dolly tub or bin,

OPPOSITE LEFT
Pumpkin flowers usually last less than 48 hours; bees visit them often in the very early hours of the morning. The female flowers fade to reveal their baby fruits at their bases.

RIGHT Pumpkin 'Crown Prince' with skin the colour of a Cream Legbar hen egg. It has been picked with a generous stalk that will add to its character as a still life ornament in the house over the winter months. Also worth growing is pumpkin 'Turk's Turban'.

two pumpkin plants can be grown. If growing a small variety of pumpkin in a window box, plant just a single plant as it will need all the soil to itself in a small container to have a chance of producing any fruit.

The growing of larger leaves will signal that the pumpkin is well. Most varieties are fast growing but not all, some take a few weeks to get into their stride and any cold nights will stall them.

Growing a pumpkin vertically up a wigwam or a trellis means that any fruits will dangle like baubles, and this helps with ripening because the light reaches all round the skin so there won't be a brown, rough mark on the fruit, which is what often happens when pumpkins are grown on the ground. Over the summer, leaves that go yellow can be cut off; the fresher ones will carry the show. Once several fruits have started to grow you want the plant's energy to go into these, so find the growing tip of the plant and cut this off to stop the pumpkin from exhausting itself.

The harvest of a pumpkin is sad in a way, as it means that autumn is almost at an end and it is time for the garden at large to be put to bed, its jungle canopies suddenly looking more brown and drooping than alive and bejewelled as the evenings draw in earlier, with close fogs.

The pumpkin should be cut with a generous amount of stalk attached; this will give it lots of life once it is taken indoors and helps it store for much longer. We place ours just like they are vases of flowers, a living Kate Malone (a potter known for her huge glass pumpkins) piece of work, a joyful globe of a being that will last and look beautiful for several months.

OVERLEAF

LEFT Morning glory, *Ipomoea purpurea* 'Grandpa Ott', on a birch wigwam, a fast-growing and worthily traditional annual climber that should be sown in early spring and planted out in late spring once it is growing well. If potted on several times until early summer, to keep them vigorous, they can be a good replacement for gone-over sweet peas.

RIGHT Tea lights in rusted stakes can be moved around the garden easily, the glass tubes offering the flame some protection from the wind. Candles create beautiful, temporary evening light, unlike obtrusive solar panels. Remember also that bats and moths don't like lots of night light! Shown here are Galdioli 'Indian Summer' and *Panicum capillare* 'Sparkling Fountain'.

Flamingo flowers

I must be careful with the colour pink because it plays around with bolder colours in the same way that white can. If it is too light or too sickly in its tone, the presence of it can be eye-aching.

Because pink comes in such an array of different but subtle tones, it is easy to plant the wrong, distracting shade of it into the Venetian-toned garden. Before you know it, pink can make the garden verge into the Barbie-doll section of Toys"R"Us, outcompeting the other colours.

The rare and correct tone of it for me is a pink that contains orange! Orange is a good colour, it is rich, alive, vibrant and happy. When orange is mixed with a deep pink it becomes the deepest of coral – not quite fuchsia and not quite vermillion orange either, but something furiously fabulous in between. The sort to be seen on the flesh of a proper wild salmon and the rich blush on the skin of a fully ripened pomegranate.

Caribbean flamingos, which would naturally live in enormous flocks on the salt lagoons around the West Indies, possess this most sought-after of flower colours in their plumage. There are six species of flamingos in the world and these are the most luscious in their colouration.

Everything about a flamingo is almost on the verge of the ridiculous, yet they are the most beautiful of nature's curiosities. These birds have fascinated me for my entire life; their colour and excitable movement is captivating. Flamingos flap maddeningly into the chaotic, they are confident thanks to their plumage yet at the same time they are fragile and forever nervous.

OPPOSITE A cross-pollinated opium poppy resembling a crazed Muppet-like pom-pom creature. True varieties of *Papaver somniferum* var. *paeoniflorum* that are available as seed include the pink 'Flemish Antique' and the sultry 'Black Peony'. Aptly, these are known as the peony poppies.

LEFT A Caribbean flamingo preening as if it were a ruffled flower on a single stem. Its other delicate leg is tucked closely underneath its body in a characteristic posture. What look like flamingos' knees to us are in fact their ankles, hence they can walk in such a crazed zigzag fashion. The slightest disturbance can cause lasting unease for these highly gregarious birds. They hatch out as grey chicks from a chalky-shelled goose-sized egg, taking three years to moult into their truly pink plumes.

I nearly became a zookeeper, having secured a work placement at Cotswold Wildlife Park in Burford, Oxfordshire, but I was then offered a year's training at Kew Gardens that same summer. Somewhat regrettably, I chose studying plants rather than flamingos, but I know a quite ridiculous amount about them, and one day I hope to go back to a more zoological than horticultural pursuit.

My favourite, little but very well-run, zoo is a bird garden in St Ives, Cornwall, called Paradise Park. Here, preciously rare toucans peek out their huge, vivid, lemon-sunset-lolly-coloured beaks teasingly from behind the creepers and fronds in their lushly manicured aviaries. The flamingos opposite them flap excitedly on the vivid green stage that is their lawn, showing off their black flight feathers and their long, curving back plumes that they fan upward. It is as if they are made from deeply serrated peony petals, like living Philip Treacy, Cher-tour-worthy hats.

The flamingos preen here to a dimming yet fabulous chorus of whistling and chattering parrots. The deep pink colour of all flamingos is diet-dependent; using their uniquely upturned, croquet-mallet beaks, they filter-feed algae and shrimps. These, once digested, give them an essential carotene pigment. In captivity, this essential rouge blusher is given to them in the form of specially made pellets that are soaked in water. A bucket of them looks like a discarded breakfast bowl of oversoaked brown Cheerios!

Flamingos do everything as a collective group, copying each other's actions with perfect choreography; they cannot exist alone. Just like a flower garden can never have enough flowers in it, a flamingo flock can never have enough flamingos within it either. Small flocks in zoos often feel calmer in aviaries that have large mirrors, as the reflections help them feel like they are part of a larger group. Feeling relaxed is essential for flamingos to flourish, as only then will they nest. Flamingo flirting involves a lot of dancing. The whole flock becomes one tightly marching symphony of head turning and ruffled wings, and so a group of dancing flamingos is known as a flamboyance. The name flamingo aptly comes from the Spanish word *flamenco*, meaning flame.

To ensure healthy flamingo feet, a soft sponge-cake-like surface that resembles a sandy lake bed is essential for them underfoot, to avoid any abrasions from developing, as a bad foot can be fatal for a bird that spends its whole life walking on stilts! The brick of the yard therefore would do flamingos no good at all, so I gave up on the fantasy of netting off the whole garden and having some of the dolly tubs filled with water a long time ago. Flamingo guano is also rather strong-smelling!

I don't want any dire plastic lawn flamingos about the place but we do have a sweet little tin flamingo watering can called Flora, who came from the now sadly closed-down Pier shop that was once full of gorgeous things quite a few years ago. Water pours out from her beak, of course. She's a character fit for a Bill and Ben, the Flower Pot Men remake, with Vanessa Feltz, I hope, as her character voice!

RIGHT A flamingo watering can enjoying a surround of taffeta-like peony tulips 'Antraciet'

ABOVE Oriental poppy 'Patty's Plum'.

Roses

So certain flowers must take the place of longed-for flamingos. First on this crazed list are roses. I like the huge-faced, *Alice in Wonderland*, old-fashioned shrub varieties that look as if their petals could suddenly begin to move and talk to you. Some newly bred sorts have the old rose characteristics of charm and scent but they come with a much-needed modern rose vigour of fast growth and disease resistance. I have several of two favourite varieties planted in a large, old, cattle trough. These are here permanently, and such a planting could be used as a large floral-like boundary hedge, the height and length of the trough helping to give immediate impact.

The most Caribbean-plumed plant of the lot is a David Austin rose called 'Summer Song'. It flowers early, from early summer, and its orange, pink-kissed punch-bowl flowers smell of bananas or, as my mum commented, like a new baby! Its partner in flower is the rose 'Hot Chocolate', an incredible rose whose burned, red-clay charged complexion changes through the day, with its flowers held on very strong stems so it never droops down like a lot of thinner-stemmed varieties do. 'Summer Song', admittedly, does droop, but staking using strong bits of birch is essential for roses and helps to prevent collapsing. This staking needs to be done in the winter, as does any pruning, when crossing stems and dead growth need to be removed with secateurs that have been cleaned between each plant. Always prune at a slight angle, down to the bulge of a promising bud or the swelling of a node.

OPPOSITE Rose 'Hot Chocolate' springing up from its anti-black-spot guardian angel, the profusely flowering *Salvia × jamensis* 'Nachtvlinder'.

ABOVE LEFT Roses 'Summer Song' and 'Hot Chocolate' cut into a glass decanter. By cutting roses you are actively encouraging fresh growth to be produced, which is helpful to the plant.

ABOVE 'Summer Song', one of the best rose faces.

Bending a rose stem over in a downward curve, once they are a few years old and having grown a good framework, encourages flowering growth much more than just leaving them to grow upward at their leisure.

Push three winter-cut stems of straight hazel about 2.5cm (1in) thick around the inner perimeter of a large circular pot. Then push each stem end down to create three half-moons that are about 30cm (1 ft) high in their middles. The long stems of the rose can then be tied down onto these, creating a loose-domed cage. The stress that this puts on the branches will cause buds to erupt en masse. This style of staking has been used mostly notably, and with great and glorious effect, in the gardens at Sissinghurst in Kent. A visit here in early spring will reveal this skill, which is quickly hidden as the garden awakens.

Why I do not just have a rose garden in all the large dolly tubs and bins, I don't know; it would be a far simpler affair. In a pot the size of a dolly tub I would plant one rose, right in its middle.

Roses can be expensive, but purchasing them in the winter as dormant bare roots makes buying a large number much cheaper than getting them as potted plants. They will need to be given a long drink overnight, then planted on a frost-free day, sprinkled with mycorrhizal fungi to help ensure healthy root growth.

BELOW A typical wooden panel fence adorned with clematis 'Etoile Violette' and good old shrub roses 'Charles de Mills' and 'Tuscany Superb'. Both of these roses can be treated as small climbers, their long stems fanned out and tied on horizontal fence wires to create floral walls and an appealing and comforting space for birds. Even better would be to include a climbing honeysuckle for its nectar-rich, scented flowers and for its berries in the autumn.

If kept healthy, roses can be spectacular in large pots, but finding annuals to plant with them can be hard as they do not like competition. You could still go to town with them in containers, replacing them with dahlias as a more permanent but still floral-rich summer and autumn garden.

Their foliage in the spring is beautiful and this element of them is often undervalued. Their fresh leaves are of a copper red and zingy green, so they complement the oranges and purples of tulips well and a good few bulbs could be planted around the base of a rose that is in a pot by using a sharp and thin-bladed trowel, working each bulb down around the root ball of the rose. The thorns always manage to stab me somehow, which results in a lot of swearing to one's self.

Roses are hungry plants, requiring several generous handfuls' worth of mulch throughout the year; rich, well-rotted muck is best or good compost mixed with organic chicken manure pellets. Add a mulch in early spring that's a good few centimetres (inches) thick, and then another to the pot's surface in late spring to help reduce water loss. These mulches will provide a good consistent source of nutrients for the roses while they are growing and in flower. Roses drink a lot; they will quickly droop in the summer's heat, demanding a bucket's worth of refreshment each evening after a hot day.

Any rose that becomes dry and undernourished in any situation, be it in the ground or in a container, will not be healthy. They will quickly develop yellow leaves and black spot, the age-old enemy of the rose garden. This blotching will take its hold on seemingly every leaf to be seen on a plant.

With black spot, like a lot of things when it comes to plants' ailments, prevention is better than cure. Watering and feeding are the main things that will stop it getting too profuse on a rose and this good care will also prevent powdery mildew, but enlisting the help of salvias as an underplanting is the most natural weapon to deploy. Not only do these not outcompete the roses at all in their growth but they look fantastic, filling out and creating dense clouds of bee-attractive flowers that bloom from mid-summer to late autumn. It is thought that their pepper-smelling foliage emits a sulphur-like quality that is released especially on hot days. Only certain salvias will work well with roses. The best and most beautiful is *Salvia × jamensis* 'Nachtvlinder', which has small, velvet purple flowers in profusion; similar to this is *Salvia microphylla* 'Dyson's Maroon'. You should prune salvias in the spring, as giving them a hair cut in winter makes them susceptible to frost and can kill them.

Remember to remove any leaves from the roses that are yellow and have fallen off, as this will further help to stop the fungal spores of any black spot lingering.

LEFT 'Goliath', the largest and most robust of the perennial oriental poppies.

RIGHT Poppy petal in the rain, like a moulted feather.

Poppies

Perhaps the most flamboyant of flowers, in the most lavish of good, deep pinks and reds, and with delicate petals of the finest taffeta, poppies burst into the garden in early summer. The only reliable poppy presence comes from the perennial *Papaver orientale*, which grow a basal clump of hairy leaves from early spring. From these rosettes appear dragon-like, frilly fat flower buds that open to reveal the most divine fairy-tale ballgown dresses. They are happiest in well-draining soil so suit pots, but I have given them pride of place in the yard's raised stone flower bed, despite them being almost as fleeting in their flowering party as peonies – they last less than a fortnight, especially if the weather is poor, when they will get painfully ruined. Staking each clump with birch or hazel before flowering begins is essential if they are to stay upright. Their saving grace is that they can be sharply cut back to the ground as soon as they finish flowering, as by then their leaves look very tatty. Annuals such as cosmos can then be closely planted around chopped-back clumps, which does not affect them at all because their fleshy roots are quite a bit below the ground's surface. A month after being cut back, fresh little leaves will sprout, and sometimes a second, odd flush of flowers will appear in the autumn.

Opium poppies, *Papaver somniferum*, live in completely nomadic, short and wild fashion. These are fast-to-flower annuals that refuse to be tamed. Freshly bought

OPPOSITE Oriental poppy 'Harlem' and the leaves of perennial *Ageratina altissima* 'Chocolate'. Although their display is brief, Oriental poppies bring the garden a rich, unrivalled carnival of taffeta.

seed must be directly sown but often the seedlings will not appear until a year or so later, once they have experienced the chill of winter. They hate being disturbed, so if a good number germinate in a pot, the plans of replanting it for summer with other purposely grown plants are put on hold and the poppies, with their silver, smooth leaves, are left to grow in it. This is worth it, though, as each seedling will send up a flower of total cross-pollinated surprises. The bees visit these flowers incredibly early in the morning as few flowers are as rich in pollen as opium poppies; the bees literally roll around in the flowers buzzing frantically.

Gladioli and crocosmia

Rocket plumes of craziness will be ensured if snobby presumptions are dropped and some select, single-coloured varieties of summer gladiolus, such as the deep, almost-black, claret-velvet 'Espresso' and the lava-like 'Magma', are included in the summer display. You do not need many of them, just half a dozen planted with the dahlias will look incredibly glamorous, spearing up through the dahlia foliage with gusto.

Gladioli are corms, and they get bigger and better with age. Each autumn, I lift them with their foliage attached. They are then shoved in a corner of Min's garage and the foliage dies down, the goodness from the leaves going back into the corm. By spring, the stored corms are big and fat and the old foliage is cut off. A strange callus often forms at the base, which I break off. They then look as ripe as newly bought ones but bigger.

A hardy corm and close relation to gladioli is the crocosmia, or montbretia; the small-spreading varieties are boring and terribly thuggish, but one of the tallest and most striking is 'Lucifer', with trumpet flowers of the truest vermilion orange, which will begin to flower in late summer. This corm is hardy and it will clump up well in pots. To ensure that flowering occurs every year its leaves must be allowed to die back, so any pots of it will need to be allowed to fade until they go brown. Fortunately, its seedheads look particularly structural in the garden if they are left over the winter. In the spring, pots of these corms should have their dead foliage cut back, then every few years the pots should be emptied and the corms split and replanted to encourage vigorous flowering rather than just a lot of leaf.

THE ORNAMENTAL WILD

Wildlife should be valued as the beating heart of the garden and yet it is still often overlooked, especially in terms of the beauty, therapy and animation that it can bring.

Why can't we make the whole garden hospitable for living creatures? That doesn't mean allowing your garden to run wild. It can be as formal as you like, but we can choose plants that are both beautiful and life-giving.

This is something I always bear in mind when I choose what to grow and how to garden, though I have a love/hate attitude when it comes to garden guests. I'd like more dashing peregrine falcons diving out of the sky; they'd sort out the feral pigeons that gather in the derelict buildings all around us. There are so many pigeons that before I take any photos of the garden, I have to scan it for their moulted feathers.

I want a Beatrix Potter watercolour idyll of twittering sparrows, singing robins and slug-eating frogs! I could even tolerate flopsy bunnies, having once had a black lionhead rabbit called Benjamin that I treated like a living pair of gloves. I'd carry him about like a muff; when I put him down, he'd immediately attempt to shag my boots with great enthusiasm! I don't suppose rabbits would bother to jump up into the dolly tubs anyway, unless they were very hungry. And if only we had hedgehogs!

The most important thing for all gardeners to do, regardless of any preference for certain species, is to totally forbid the use of chemicals. Herbicides and pesticides have somehow become cunningly embedded in many gardening tasks; the aisles of garden retailers are full of them. They are still seen as the easy solution, force-fed to us, even in some professional horticultural training. But thankfully attitudes are changing – a few years ago you'd have been considered an over-sensitive hippy for shunning pesticides, now the science is telling us that such chemicals are not just bad for insects, but for everything on the planet, including us! If we could banish these chemicals, then, as gardeners, we could do a huge amount for biodiversity. The pesticide industry has vast power and influence within the domestic gardening sector. Make yourself aware of these mega agrochemical companies and stop buying their products!

Just one application of pesticides, particularly from a broad-spectrum group of chemicals known as neonicotinoids, is arguably harmful, especially to our vital and beautiful bees, the angels of this Earth. The plants absorb these chemicals into their entire vascular systems and so their nectar and pollen are then laced with fatal poison.

The effects of this crop treatment to honey bees in particular has been devastating, resulting in the term known as colony collapse, where entire hives of bees die out within days, because the little minds of the bees are so damaged from these chemicals that they can no longer navigate. Add to this huge monoculture crops, notably rapeseed, supplanting flower diversity and local councils still being often reluctant to turn parkland and roadside verges into proper flower meadows, meaning that bees really need our gardens as refuges.

The plant industry has, to its credit, become more aware of the damage that neonicotinoids do and many retailers have committed to stop using them, but it is worth asking questions when you buy plants just as you would do perhaps when asking if the custard in a restaurant has been made with free-range eggs. Supporting small, independent nurseries, which are more likely to be taking an organic, non-chemical approach from the outset, will usually be the safest way of stocking your garden with truly wildlife-friendly plants. Most seed merchants should be able to confirm what chemicals their growers are using in the growing fields and whether they are bee-friendly. But again, it's always worth asking. The organic bulb market is now growing, too, which is very encouraging.

BELOW 'Tuscany Superb', a good old-fashioned rose. This bumblebee's pollen baskets are full, thanks to this rose's generous pollen-laden anthers.

ABOVE Bumble bees in opium poppy, somniferum 'Lauren's Grape'.

RIGHT Millet and panicum, annual grasses creating a jungle tapestry with dahlia 'Totally Tangerine', a feast for bees and birds.

Songbirds

Mixed dried fruits, fruit loaf, overnight-soaked mealworms and chopped apple make up the daily menu that my mum serves to Mr and Mrs Blackbird.

The blackbirds find sanctuary in our tiny back garden, which is Mum's domain. It's smaller than the front garden and is almost completely shaded by a huge neighbouring horse chestnut tree, so many of my favourite flowers do poorly here as they are sun lovers.

I try not to be too much of a cuckoo in my mum's patch. I admire her patience in letting plants fade away naturally, rather than my own 'off with its head' approach.

In place of sun-loving annuals are salvias that cope in the dappled shade. They include 'Amistad', perhaps the most popular of the lot, 'Love and Wishes', whose vibrant pink petals appear from burgundy slipper-like bracts, and the lava-like 'Ember's Wish'. Altogether these create a paintbox of primary colours through summer and autumn in the little stone flower beds.

If the winters are mild and these salvias are not pruned back until spring properly arrives, then the majority of them will prove to be perennial; a compost mulch around them will help. More reliably hardy, though, are my mum's spike lobelias, which flower in a range of purples and maroons. Perennial *Phlox* 'Blue Paradise' does well here, too; if you can't be bothered to grow sweet peas, grow this instead for its perfume.

Unlike the front garden, the back garden is surrounded by a fence, so it is a proper shoebox with sides and a back. The roses 'Charles de Mills' and 'Tuscany Superb' climb up the garden's fence, their stems tied onto wire supports each winter. They are planted into the ground and married to the *Clematis* 'Étoile Violette' that snakes up through them. Clematis need to be fed and watered a lot, especially if they are in pots, and they will only thrive in the largest of these. A roofing slate should always be shoved in at the base of newly planted clematis so that they are shaded from the sun.

Try to use climbers to dress the fence, like garden walls; if you have a new garden always plant these first, as the most ornamental and beautiful of them take time to get going. People are frightened of the fast growers *Vitis coignetiae*, the crimson glory vine with its dramatic leaves, and its relation, the country-pile-cladding *Parthenocissus quinquefolia*, or Virginia creeper, because they grow so fast. I have had to spend hours with a paint scraper removing tightly clinging Virginia creeper from brickwork, but to see these climbers in the autumn fading into burnished reds, maroons and oranges is quite a sight, so I think it's worth having

PAGES 94–5 Tin baths in midsummer. Dahlia 'Waltzing Mathilda' covered in bumble bees, with chocolate cosmos and the tendrils of the soon to flower Thunbergia alata 'African Sunset' clambering through.

PAGES 96 A honeybee (top) and a bumble bee (bottom) sharing the nectar and pollen of dwarf sunflower 'Ms Mars' – late sown annuals that flower between early and late autumn. For a shady garden, the annual Nicotiana or tobacco plants will flower happily in such places and their night-time scent will be enjoyed by moths.

OPPOSITE The ornamental wild – Mrs Blackbird visiting for lunch in the calm of the back garden.

them, if you dare. If you plant something vigorous and prune it often, you will keep it under control.

I love hazel hurdles and I would definitely save up for them if I needed a permanent fence, as they instantly give a lovely cottage-garden feeling to any space. Having a smothering climber, like a Virginia creeper, would help hold them together once they go brittle and start to loosen and eventually fall apart. I am tempted to suggest planting them with ivy for the same reason, but this may be risky – the runners that ivy puts out into the ground would need to be kept firmly in check. Ivy, however, when it is in flower, is one of the best blooms for pollinators, so don't be too quick to cut it all away; recent studies have also shown that it helps to insulate stonework and buildings rather than destroy them, but obviously keep it from reaching any roof tiles!

You must take the time to paint any hazel or willow hurdles with a good wood preserver, as it will prolong their life for several years. The hurdles make very good pens to hide dustbins, too!

A garden with fences covered in foliage will not only feel wonderfully engulfing but will be more accommodating to visiting birds; the blackbirds certainly seem to feel happy in the back garden, a refuge from a hostile world, with more cats, more magpies and fewer hedges than ever. Town life, indeed any life, would be unbearable without waking up to the defiantly triumphant song of the male blackbird; in the summer, toward the longest day, they sing long into the evening. Heavenly. But how blackbirds are managing to find enough worms and bugs during the almost rainless Aprils we have had over the past few years amazes me. They are particularly grateful for the soaked mealworms they are served in the morning; soaking transforms them from dried crisps into soft delicacies that a baby blackbird can easily swallow. These treats have to be hidden between pots so that they are not scoffed by greedy wood pigeons, though.

The clacking of magpies sends shudders through me and Mum – we hate them, and with good reason as, working as a team, they will cunningly distract a blackbird's attention away from the nest. They are highly intelligent birds and once they have discovered a blackbird's nest, it is doomed.

There are more and more magpies now, especially in towns as they have adapted their diets to include litter as well as baby birds and eggs. Our blackbirds are under siege and we can do nothing apart from try to keep the parents well fed enough to nest again. Luckily, a pair of blackbirds can rear as many as three broods of chicks over the spring and summer.

ABOVE A glazed pot of Moroccan blue complements the orange lanterns of Physalis alkekengi, the perennial Chinese lantern. The lanterns are cut in late autumn to dry inside before the weather ruins them. They keep their orange pigment well for months in the vase.

RIGHT My mum's domain, the back garden, is full of treasures. In early Autumn, its primary-colour tones come into their own, thanks to a pot of smile-inducing Chinese lanterns. Salvias and lobelias thrive in the dappled shade, as does perennial phlox 'Blue Paradise'.

LEFT My red factor canaries enjoying millet sprays, just as wild birds do. Many former mining towns like Nottingham would once have been full of canary fanciers. Poor fliers, and entirely domesticated, they are content in large cages.

RIGHT A bird banquet. Millets *Setaria italica* 'Red Jewel' and *Panicum violaceum* with *Amaranthus cruentus* 'Hot Biscuits'. All these annuals provide invaluable foliage for dahlias and cosmos and if you store the seedheads as they ripen over the summer you can put them out over the winter in bunches as chic bird feeders.

FEEDING BIRDS

Bird feeders are a difficult subject. Unsurprisingly, I can't stand the sight of most of them and the best ones, those that are designed to stop grey squirrels from feasting, command huge sums of money and yet still look like something designed to be put on the moon! All feeders must be taken apart and washed at least once a month in hot soapy water with some white vinegar, otherwise you risk giving the bird diners aspergillosis and avian pox.

But feeding garden birds is now arguably a lifeline for many. In the case of goldfinches, the feeding of niger seed has boosted their numbers hugely over the past decade. They have gone from being a shy country bird, once trapped in the thousands to fuel the Victorian caged-bird trade, to being a regular visitor to suburban gardens.

The answer to feeding garden birds decoratively lies in growing your own millet from seed. Forget the frumpy bulrush type that you might have seen in park bedding schemes, the more refined and cheaper *Panicum violaceum* is the ultimate ornamental annual grass and the bird food of choice. Teasels are good, too, but they are thuggish in comparison.

Sown from late spring to early summer, in little batches, the millet germinates in less than a week and grows very quickly. They are a perfect supporting cast for dahlias and cosmos. The seedheads burst out from the stems' middles to reveal a candelabra cascade of a thousand marbled green and purple seeds that ripen to

OPPOSITE

TOP A cock sparrow in the garden. The wigwams of sweet peas form temporary hedge-like cover and birds such as sparrows find shelter within them while also pecking up a meal of aphids.

BOTTOM Red millet, the best bird feeder. Initially green, it ripens to a marbled purple.

gold. They can either be left for the birds to find them or they can be picked and stored (hang them up somewhere dry and airy), then put out in winter as a hanging bunch. Blue tits and finches will especially relish them, as will chickens, which will have to jump to reach them. If I had an allotment or a large garden, I would grow great beds of this millet. My red factor canaries love the seedheads; it is as if I have put a canary version of caviar into their cage!

It is not just seeds, however, but insects that fuel many garden bird populations. A lot of small birds live short lives, so a crash in insect numbers for just one or two seasons can have a huge impact. House sparrows have been hit particularly hard,
but here every summer pairs of them visit not the back, but the front garden.

What brings them is not seed but aphids – greenfly gathering on the tips of the sweet peas and roses. For several weeks the yard is alive with the flitting and chattering of sparrows as each pair takes it in turn on bug patrol duty. The sweet pea birch wigwams form a floral hedge, where they feel safe and at home, able to come and go as they please, back to a loose roof tile on a neighbouring building. So, garden birds that need insects to feed their chicks are the most natural aphid control you could ask for.

Perennial wigwams of honeysuckles would be good, too. The berries of these might even attract my favourite songbird of all, the bullfinch. They are very shy, so I had better up my planting game if I ever hope to see one from the kitchen window while washing up.

HEDGES

I am sad not to have hedges. Cover is the most vital thing for any garden to truly provide sanctuary for birds; they particularly love hawthorn. If you have hawthorn hedges, birds ping-pong across a garden to and from them. The older and thicker the hedge, the more attractive to birds it will be.

There is a lot of talk about planting more trees to create more forests, but no one seems to be advocating a mass planting of native hedgerows. Trimmed twice a year, in late summer and late autumn, to keep them neat and low, hawthorn hedge-lined streets and pavements, especially in new-build housing estates, would create beautifully domed lime-green avenues that would be alive with birds.

I would certainly have hedges if I could plant them into the ground, but I can't, so I need alternatives. Crab apples on dwarf rootstock are happy in a large dolly tub, providing blossom and, in the case of some, such as *Malus* 'Royalty', beautiful crimson leaves. Crab apples hold their fruit long into the winter months and once several frosts have decayed them a little the birds feast on their branches. Bushes of spindle berry 'Red Cascade' are worth planting, too, for their autumn splendour, although they are a little dull until then.

Bees and butterflies

There is nothing more magical than seeing bees and butterflies visiting your flowers. Even in a small space you should try to provide as much flower power as you can for pollinators through the spring, summer and autumn seasons. These vital insects give the garden a true sense of being alive; a window box can support them if the right flowers – rich in both pollen and nectar – are being grown in it. Think of the golden age of cottage garden plant tapestries having an abundance of flower faces, open and full of nectar, in sizes large and small, a bloom for every bee and butterfly.

In this book, almost everything photographed is pollinator attractive. While native species of wildflowers are indeed excellent from the point of view of a bee or a butterfly, you honestly don't need to turn your garden into a wildflower meadow for it to be a pollinator paradise. Garden flowers, if they are single or semi-double varieties, are hugely valuable to pollinators. These are the flowers that will naturally have nectar- and pollen-rich anthers and stamens. As a rule, any flower that resembles a daisy, with an outline of petals around its middle, is attractive to insects, as are almost all flowers that look as if they are related to thistles.

It is worth remembering that each different species of bee has its own length of proboscis (tongue), so certain flowers will attract particular species more than others. A dainty chocolate cosmos flower, for example, will hold the weight of a honey bee perfectly, but it will not easily accommodate the heavier, helicopter-like bumble bee, which will make them sway about too much. Bumbles prefer the larger landing pads of sunflowers and single dahlias.

The bottom fat lip of a snapdragon will require a bee to use a lot of effort to prise it open and only a particularly strong or large individual will succeed in doing so. Interestingly, a number of salvia cultivars, whose flower designs hail back to their ancestors growing in native jungles where hummingbirds are the pollinators, still prove useful, because a number of bees have learned to spear a hole through the tubular flowers to reach the nectar, a bee behaviour called robbing.

I love the fairy wings that bees have and their fur coats, ranging from burnished orange to jet black, to Princeton apricot-brown. Watching these creatures go about their day is essential therapy for me.

If I chose to fill the garden with double pom-pom dahlias and sponge-like trailing double begonias (God forbid), then it would be a sad, lifeless place, because, packed with so many petals, these flowers have no nectar and are hence useless for bees. I have stood in many gardens brimming with decorative dahlias, yet they are horribly silent. Beautiful but soulless. I'm not saying you cannot have such flowers

ABOVE Snapdragon 'Liberty Crimson'. A diversity of flower shapes ensures different bee species are provided for during summer and autumn.

OPPOSITE In the sun, buddleia 'Buzz Magenta' is magically alive with butterflies.

LEFT Bumble
bees visiting
*Nectaroscordum
siculum* with *Anchusa*
'Loddon Royalist'

RIGHT

TOP LEFT Bumblebee
upon *Salvia* 'Amistad'.

TOP RIGHT Martagon
lily 'Sunny Morning'.

BOTTOM LEFT
Wallflower 'Vulcan'.
All wallflowers are
especially good for
spring nectar – the
bees can smell their
perfume as well as
we can.

BOTTOM RIGHT
The tall and perennial
Aster novae-angliae
'Violetta'. This is
worth having in
pots at large. If you
cannot be bothered
with growing annuals
or dahlias and want
some rich colour
in the autumn, pair
it with the deep pink
aster 'September Ruby'.

– I love a big party dahlia – but don't let them rule the roost. Garden flowers play a big role in supporting bees all the year round but especially in early spring, late summer and autumn, when orchards are not in blossom.

It seems hypocritical of me to proclaim myself as a saviour of insects when my garden is without stinging nettles! I could have a large trough planted with clumps of nettles and if I had a larger garden or an allotment I would definitely have a big patch of them because they provide a vital nursery for the caterpillars of the prettiest of butterflies – the Peacocks, Tortoiseshells, Red Admirals and Commas. No nettles means no caterpillars and no butterflies. Nettles grow in my dad's garden, and when the black Peacock caterpillars are spotted, the entire plant is covered over with horticultural fleece to try to save them from being eaten by wasps! In America, there is an urgent campaign for gardeners to plant milk weed, as this is what the caterpillars of their iconic migratory butterfly, the Monarch, feed on.

ABOVE Dahlia 'Waltzing Mathilda' with a Red Admiral butterfly. In medieval England the admiral's colouration meant it was regarded as the devil's fly. Fallen and rotting apples will also be appreciated by butterflies in the autumn.

Insect pests and organic controls

As gardeners we do need to control insects for the sake of our plants, as some bugs can really cause havoc, but we can tackle these particular pests in ways that are effective and organic. When I was little, I remember vividly being caught by a horrid dinner lady as I tried to take a jet black rain beetle that I had found scuttling on the floor of the school hall outside into the playground. She knocked it from my little cupped hands and stamped on it with such nastiness. I hope with the passing of time that she turns into one herself!

VINE WEEVIL

The traditional menace to the container gardener and not one you can afford to ignore. What purpose vine weevils serve in the food chain I cannot understand, as nothing seems to benefit from them and nothing, annoyingly, seems to eat them – at least not in large enough numbers to control them. So action has to be taken to prevent them from quite literally eating your garden to death.

The adult, grey, tank-like beetles ruin flowers and leaves, nibbling around their edges, often under the cover of darkness. They then lay small, creamy yellow eggs in the pot's soil. These hatch into vile grubs that eat the roots of plants until they pupate into adults. This is one of the reasons why I empty the pots between spring and summer, as it gives me an opportunity to check for any vine weevil activity; it is often when pots aren't emptied out that vine weevils really take hold. Certain plants, namely auriculas, primulas and heucheras, are particularly attractive to vine weevil.

The only organic and effective solution that I can recommend is nematodes. These cultures will kill any eggs and grubs in the soil that you cannot see; you will need to collect and crush adults by hand, as they come out at night! Nematodes occur naturally and prey on their host insects when the soil is warm, especially in summer. By applying these via the watering can, you are just bumping up their numbers to a concentration that will lead to complete eradication; these are species specific to vine weevils – other bugs and wildlife are not affected by them at all. They can be ordered from specialist suppliers, and are sent out at the times of year when they can be applied, or you may be able to buy them at good garden centres.

Nematodes are alive, so they have to be kept in the fridge and applied within a few weeks of being bought. They look like a packet of yeast and are added to a measured bucket of water, which is then diluted into a filled watering can before being quickly sprinkled all over your pots, either at dusk or dawn, in a relay, until all used up.

SLUGS AND SNAILS

Visually slugs and snails cause the most damage to a garden, especially one grown in pots. The most effective treatment is again nematodes. Please don't use slug pellets; these are not effective and have made the mistle thrush almost extinct in urban areas. If you are a pellet addict, please consider switching to an organic alternative of ferric phosphate known as Sluggo. The best way to deter slugs and snails is to reduce their hiding places during the day. These are often the bottoms of pots, so for those that are light enough to be easily tilted onto their sides, a sweep underneath with a brush every few days will dislodge them. They can be collected during summer nights using shallow bowls filled with beer – they are attracted to the yeast smell – but I don't drink beer, so I usually use large oranges or grapefruits, cut in half with their insides scooped out, then turned upside down as bait.

What to do with the blighters once they are collected, though, is a question indeed. You cannot just throw them over the fence as they will find their way back, pronto! They can be dropped into a jam jar filled with salt, then once dead thrown heartlessly away. If you know someone who keeps ducks, or if you are near a park or canal, you can feed them to the mallards, who will go bonkers for them! In spring, though it is admittedly more romantic than practical, I borrow a pair of beautiful green velvet Cayuga ducks and give them a daily hour of slug and snail patrol.

Mulches around newly planted plants, or emerging crowns, with generous handfuls of sharp grit, eggshells, dried seaweed, wool pellets or mineralised straw, will help create defensive moats around the inner perimeter of the pot. I have not found copper rings especially reliable; Vaseline smeared generously all the way around a pot's outer edge does a much better job.

APHIDS

People get in a panic about black fly and greenfly, which usually gather around unopened flower buds. Ants, being ever clever herdsmen of other bugs, are often seen accompanying aphids, drinking their honeydew sap. If you notice small colonies, gently use a wet wipe on the affected area of the plant, which will remove many of them in one go. Aphids that appear on windowsills may be a sign that plant hygiene is lacking, so clean the windowsill and air any plants on it for a few hours outside.

You can apply a non-toxic spray product known as SB plant invigorator to small seedlings and plug plants, and to indoor growing situations, like a windowsill. It will also help prevent powdery mildew. Horticultural soft soap can be used too, but only on dull days as it can burn the foliage. A large number of aphids can be dislodged simply by using a hand-held water-only spray. Encouraging birds and ladybirds that eat aphids is the best approach, though, and a generally scorned insect, the menacing-looking earwig, actually eats aphids by the ton, so don't kill them unless you have them in numbers so great they are consuming too many flowers during the night!

RIGHT A Cayuga duck peering through the salvia 'Love and Wishes'. This duck and her mate live an idyllic life on my friend Jac's farm in the Cotswolds. I took them there via train, a somewhat embarrassing journey as, unlike hens, ducks are quite vocal when they are in boxes. The conductor gave me several odd looks but was kind enough to not ask any questions! The romance of marrying farmyard to garden means you must be able to cope with the mess and have a commitment to the welfare of livestock. Everything is period drama picturesque in the summer but the winter months are another matter. A line spoken by Niamh Cusack when she played Beatrix Potter stays with me from childhood, a scene of the writer running through her cottage garden in a storm to find her ducks bathing in puddles by the door: 'Out of my way please Jemima! You might be enjoying this rain, but I should like to get to my door, if you please.'

The Minnie Brown moth

A garden could be the fountain of youth in this life, decay can be cut off and young growth stimulated.

We humans, of course, have not got this luxury; there's no cutting back for us, Botox is an option, but it's overrated, and gardening is not good for the body. You don't become a gladiator through doing it (obviously), you get bad skin and dirty nails that require hours of soaking to get properly clean and you cannot garden properly in gloves, so wearing them is out of the question! The thing about gardening, though, is that when you are doing it, it fully consumes the mind and self-awareness disappears; you are part of a Wonderland-like dream in that moment. It is totally transporting, a high of earth and colour and you rediscover a childlike, carefree ignorance, but there is no comedown of any sense, only a visual reward and feeling of nurturing.

Min or Minnie Florence Brown, my nana, is my dad's mum. She lives up the road from the yard and is ninety-five; I joke that she is too old and I am too crazy. What Min gives me is consistency in company; she is almost always to be found in her bed like a little half-asleep moth, with the fat beak lips of a fledging baby blue tit chick hungry for a rice pudding with a spoonful of maple syrup or a full English fry-up: mushrooms, fried bread and all the trimmings.

Min allows me to indulge my passion for keeping poultry in her garden, something that I have done here since I was seven. 'You don't mind the chickens, do you, Nana?' I say. 'No, duck, why should I care?' 'Well that is fabulous then,' is my reply. 'Are there any eggs today, have they laid?' she asks. She likes to know that the hens are laying and her face lights up when I bring the warm, newly laid eggs to show her.

To get Min into the garden on sunny days, I must laugh her into her wheelchair, otherwise it would be a constant 'Oh no duck, another day'. We have a largely 'duck'-dominated language – 'Ok, duck; yes, duck; thank you, duck; come on then, duck; cover me up, duck.' I wheel her up the garden path and we sit beneath the apple trees on the increasingly ground elder-infested lawn, in our own little world. It is very precious time for us now; dementia like the roots of ground elder will take my nana Min from me in the seasons to come but we are not there yet; luckily she is a tough spirit.

Min was the youngest of nine children who grew up in a small terraced house. She became her mining father's favourite thanks to her being willing to water his seven large greenhouses each evening after school. They were built on a smallholding at the end of the road. Here he grew tomatoes and chrysanthemums; bunches of them would be sold down the pub for eight old pence (less than 4p)! Pigs were kept too and Min does a fond impression of her mother telling the hungry large white sow to 'get back, go on get back' as they poured the swill into the trough.

A well had been made using pit gunpowder and Min would pump the water up from it, standing on a wooden box. Her biggest school achievement was growing the best hyacinth bulb in her class, a scented mass of lilac blue. Now she has completely lost her sense of smell.

ABOVE LEFT Hawthorn hedges in their fresh lime green. Hens appreciate the shelter of hedges and the dry soil for dust baths. Gardens that have such hedges as boundaries will be alive with songbirds.

ABOVE RIGHT Bantam eggs and the tulip 'Estella Rijnveld', which is marvellous once cut but a horror if left in the garden, being of the most manmade white and red colouration.

A PASSION FOR CHICKENS

Today, my hens have given a new sense of purpose to Min's long, former vegetable garden. I have a dozen bantam hens and, given the tiny space of our yard, they are much better off here. Hens love to have a lawn to peck on – grass is what makes for a rich egg yolk, along with calendula petals. The hens live in a summer house-styled penthouse coop that is placed on predator- and vermin-deterrent footings of 1970s crazy paving.

The roofed-over run keeps out the wind and rain and the hens are content scratching about within on a thick bed of chopped-up straw, but most days they have the liberty of the garden, bringing it wonderfully alive. Chickens certainly recognise their keeper and they will identify you as being part of their flock.

They are uniquely beautiful and charming creatures; I would never want to be without them and have only been so once, which caused me incredible misery. I love the routine of poultry fancying and talking to hens. You spend a lot of time scrubbing drinkers and perches as a hen wife or husband, and I use a Henry vacuum cleaner to clean out the nesting boxes; it works a treat. In the winter, the hens' feathered feet often need to be washed too!

My steadiest bantams of the flock join me at Mill Yard for an afternoon's gardening. I carry them down the road in a light but roomy plastic plant pot covered over with a metal soil riddle that fits tightly. They are duly released by the door and they set about bug patrol, weeding in between the bricks with their beaks as I potter about, too. When I lift the pots up, they dive for anything that crawls.

If I turn my back, though, they soon tootle out into the yard and must be scooped up. Bantams are lovely and better for a small garden as they are gentler with the plants, but I do miss collecting proper, big eggs.

Hens seem to become calmer with age and their personalities develop as time passes. Surprisingly, a garden hen of pure-breed lineage can often live for five or seven years. Hybrids, because they have been bred to lay their hearts out, live less long, usually about four years. Bantams keep laying into their pension years and always keep themselves looking neat because they give themselves lots of time off from laying, so they are rather sensible really! Little eggs have surprisingly large and creamy yolks, like superior quail eggs.

Fresh eggs encourage even me, someone with little interest in food, to cook. Last night, Min and I had poached eggs and asparagus. Luckily for me, Delia Smith has some good online cooking tutorials and the newly laid eggs held on to their whites perfectly in the vortex of boiling water.

Each spring, I pot up and grow the dahlias here. If I did not have this space, I would have to find an old greenhouse for dahlia growing. Unlike the neighbouring gardens, Min's is still bordered by hedges, lime-green bird cathedrals of hawthorn. There is a fallen-down pear tree that drapes itself elegantly and acts as a parasol for

ABOVE A frizzle-plumed Buff Pekin finding delight in the seeds of panicum 'Sparkling Fountain'.

OPPOSITE A pigeon-sized and very talkative little Belgian Barbu d'Uccle Millefleur bantam hen. The d'uccle is in reference to this breed's beard-like plumes under their beaks which often need to be washed! She is perched on woven hazel that is supporting a clump of peonies 'Rubra Plena'. Compare this freeness of look to the stright-jacketed, controlled effect that a smoker's teeth-coloured bamboo cane would have.

the hens on sunny days. There is a beauty to an almost wild garden that is magical and refreshingly relaxed.

The lavender bushes are old and leggy, but when they flower, they are stunning, great grey wigs, hazed with lilac-blue. I prune them back each spring, very lightly. Marjoram has self-seeded all over the garden and when it's in bloom it's a razzmatazz of pink and purple. When this and the lavender are out the garden becomes a haven for butterflies and bees. A great square that was once a huge bed for vegetables is now a carpet of oxeye daisies, the result of Min and I sowing a mix of perennial wildflower seeds some years ago.

Perhaps this is the right time to note that herbs are the most undemanding plants to grow in pots and will largely thrive on neglect, with just some seasonal trimming to keep them bushy.

The roses in Min's garden have lasted well. These are not the thin-stemmed modern cultivars but beefy, town-market tea roses. Their big flowers are held on

ABOVE LEFT Lavender and marjoram, a heavenly wild style for insects of all species. Such a planting would suit a perennial herb pot scheme too.

ABOVE RIGHT Marjoram 'Herrenhausen' is a particularly good variety, but all of them are beloved of bees and butterflies.

strong, upright stems, ideal for cutting. They get pruned quite harshly each winter and up they come again with renewed vigour in the spring.

A large clump of beetroot-red peonies, decades old, hold court for a brief fortnight. Many people new to gardening declare they want to grow nothing but peonies, without realising you get a mere two weeks' worth of flowers each year. I keep meaning to divide the peonies, which you can do, but only in early autumn when they do not mind being disturbed. The key to ensure that they flower well is never planting their crowns more than 2.5cm (1in) deep under the soil.

This wildness is in total contrast to the control and order once found in Min's garden. For her generation, the vegetable bed was the larder – it had to produce! Then the garden was full of beds containing leeks and purple sprouting broccoli, planted in neat rows. Min liked her roses and lupins, but there was no acceptance of insect life, as proved by the cupboard of death in the garage, its shelves packed with poison, weed killers and bug spray all decanted into recycled bottles with handwritten labels. Nan's generation were taught and encouraged to be masters of the garden through biocide; everything had to be clipped and controlled; bugs were to be feared and killed. They did, however, look after things; Nan would turn old clothes into netting to protect her raspberries and any holes in these clothes would be stitched up by hand with huge care.

In the neighbouring gardens to Min's, as each elderly owner has died, these once actively tended plots have been fenced rather than hedged, paved and lawned. Not one of them has a flower bed. They are almost identical: a uniform of patio, mown grass, trampoline and enormous shed. When the sun appears, out come the mowers and barbecues, like clockwork, the air fills with smoke and noise. These gardens are sterile, almost dead spaces that create a frightening view of disconnection to the natural world.

Ernie, who used to live next door, had a total wilderness that swamped his huge apple trees by midsummer. And yet every autumn his trees bore more apples than you could count. Their branches, full of canker, would be weighed down with fruit. Min would tacitly encourage me and my brother Lyndon to go scrumping. We would push each other over the hedge like Peter and Benjamin Rabbit, watching out for our own Mr McGregor, who would loom up from his back window if he noticed the tops of the nettle clumps swaying oddly as we crawled to the trees with our buckets. Min would make apple juice with the scrumped apples; it was like nectar with a froth. Ernie would rather have seen every one of his apples rot on the ground than have known that any had been picked and enjoyed; he was like an old blue-faced turkey stag of a man in the end but at least he didn't cut the trees down and the blackbirds feasted on the rotting fruit all winter.

BELOW Min's cupboard of endless and historic products of garden death, which would make a sinister *Midsomer Murders* cocktail of dark desires! Such poisons have no place in a modern garden or world.

MOVING INSIDE

We use the top of a shoulder-high, flat-topped wall that divides the kitchen from the living room to display flowers from the garden. It is often the only free surface, and forget the in-vogue country house term 'flower room' – this is a cottage after all – preparation happens at the kitchen sink! So, this wall is constantly covered in vases, a shelf of flowers through the year. When the sun shines through and hits the water-filled glass, they all sparkle and shimmer like a stained-glass window.

Just like arranging pots of plants on a table outside, with the largest one in its middle acting as the heart and the other smaller ones around it like a bubbling table runner, the same sort of order can be given to cut flowers inside. Several vases with a riot of flowers within them often work best. Sometimes, however, just one flower such as a huge dahlia, a furiously ruffled gladioli or a rose cut on a long, arching stem can steal a room as a single act, if it is styled in a stately vase such as an old glass decanter.

Flowers cut from the garden are powerful inside; they can transform sad shells of rooms and be the crowning glory of those that have been given the most lavish and expensive interior attention. Scent is an important element of cut flowers, too, not just looks. Sweet peas are the obvious perfume candidate for early summer but before them spring bulbs, such as hyacinths and narcissus 'Pheasant's Eye' (the only daffodil that I like to grow) with its spicy, cinnamon fragrance, echoes that of wallflowers and, like them, is intoxicating.

Cutting flowers

The yard is not a cutting garden in a slash-and-burn daily style possible if it were larger, but this does not bother me. Great consuming bundles of flowers fanned out so that they eclipse their grower, made famous of late by the American field flower-growing market, need to be forgotten about. This is not like growing a crop, but rather growing a garden.

More for the sake of more is often not a good thing. Flowers grown by the door are to be seen as individual characters rather than as a collective. Plants are rarely grown in rows here, but they flower instead in their colony-like islands within each pot.

The small garden can still be productive but in a far more intimate and appreciable way. Flowers I pick by the half- to full-dozen handfuls two times a week, and even this meagre harvest will prove to be enough when it is decanted into single-stem vases to demand attention and add beauty to a room.

People often seem afraid of cutting flowers, which is sad. While the term 'flowers for cutting' has made something of a comeback, there is often a tepidness to cut anything from a garden that is giving a pleasing picture. It important to overcome this for the sake of prolonging such a scene's existence. Cosmos, for example, will thrive on being cut weekly, as will dahlias and, of course, sweet peas.

A cosmos will become stronger and bushier as a result of cutting; vigour is installed into the plant after each harvest, so they will flower for far longer into the autumn. Each time a flower and its stem are removed by cutting down to a pair of leaves, this tells the plant to produce new growth that will result in more flowers.

In comparison, a cosmos that is not cut but is instead traditionally dead-headed will usually stall noticeably sooner. Unsurprisingly, dead-heading is a task that I don't need to do too much of because most flowers are live-headed (cut for the vase) and yet the garden almost always still looks full of flowers because cutting ensures that the annual plants continually produce them in abundance throughout the summer.

Of course, spring-flowering bulbs are not cut and come again; once a tulip is cut, that's part of a pot's display removed. This is one of the reasons why I plant quite a lot of tulip bulbs, because I know that I will probably pick half of the bulbs that I plant over the spring season.

PAGES 120–1 Flight of the parrots. Parrot Tulips in single stem vases: T. 'Parrot King', T. 'Black Parrot', T. 'Amazing Parrot' and T. 'Negrita Parrot'. The weight each flower has requires a vase with a heavy base. Tulips continue to grow once they are cut, growing necks like twisting swans.

PAGE 122 Hyacinths 'Woodstock', 'Jan boss' and 'Anastasia'.

Vases

If cut-flower harvests that consist of a dozen or less flowers are to have clout, each flower needs to retain its own identity and to be allowed to breathe once they are in the house. I have long been a fan of using single-stem vases for this sort of look rather than being a slave to stem-eating bigger vases.

We have one beautiful, old, green-tinted glass jug that is used when we have lots of flowers to cut en masse. It is a charity shop find of my mum's that is the ideal big flower vase because it has a perfectly rounded middle and a gently curving-outward neck. This shape will ensure a romantic and free look to cut flowers, but most of

BELOW A summer carnival of bud vases with dahlias 'Totally Tangerine', 'Bishop of Llandaff', 'Darkarin', 'Happy Single Kiss' and cosmos 'Rubenza' and 'Double Cranberry Click'.

TOP LEFT TO RIGHT Calendula 'Neon' with Borage, which can be sown direct each spring and autumn. Dahlias 'David Howard' and 'Thomas Edison' in Bristol blue glass and old medicine bottles. Dahlias 'Waltzing Mathilda', 'Schipper's Bronze', 'David Howard', 'Bishop of Auckland', 'Totally Tangerine', 'Bishop of Llandaff' and chocolate cosmos. Gladioli 'Plum Tart', 'Magma', 'Bimbo' and 'Espresso'.

MIDDLE RIGHT Sweet peas 'Black Knight', 'Matucana', 'King Edward VII' and 'Blue Velvet'.

BOTTOM LEFT TO RIGHT Tulips 'Orange Princess', 'Irene Parrot', 'Brown Sugar', 'Ballerina' and 'Sarah Raven' in vintage bud vases. Tulips 'Slawa' and the species 'Whittallii' join the former bud vase varieties in a jam jar bunch with snake's head fritillaries. The Rose 'Lady Emma Hamilton' in a treacle tin with viola 'Tiger Eye Red'. Ball dahlias, these last the best of all dahlias in the vase: D. 'Happy Halloween', D. 'Cornel Bronze' and D. 'Tirza' with the cut tendrils of Calibrachoa 'Can-Can Terracotta'.

the time we pick flowers into single vases, which when grouped together gives a surprising presence as a collective – much as just one large vase would. In the case of some dahlias and parrot tulips, just one great flower can have enough presence to demand attention in a room, from even the most florally dismissive of eyes.

My rules for enjoying cut flowers are to not really arrange them but to snip and plonk. It is a ritual that should be fast and fun. Fill the vases with water up to their brims and keep them topped up like this every day using a little jug or even a water baster.

You should be picking your flowers straight into water, too, as soon as you cut them; they need to be kept hydrated. Do the washing up before you go and out pick your flowers so that you have the sink and draining board nice and clear to arrange your harvest on.

A second rule is to strip off any foliage that will be submerged, as this will rot and poison the water within a day and as a lot of the vases that I use are clear glass, stripped stems always look better. I do not use flower food, but I will refresh the water completely if it is looking murky.

Each vessel in which a single stem is placed is different. A key consideration is the vases' water weight. A vase to support a tall stem with a heavy bloom on its end will obviously need to hold enough water to prevent it toppling over.

Some vases are slim and tall, others curvy, dinky and fat. Some are simply glass jars; those that you get at Christmas with liquor and peaches in often are of beautiful shapes. Catwalk-necked but Buddha-bellied decanters are a particularly good water-holding weight for a heavy single head.

The watertight tins of Abram Lyle & Sons Golden Syrup look surprisingly regal shimmering in gold and emerald green, too beautiful to think of putting out for recycling, so we have lots of these.

Clear and boldly coloured glass vases, old medicine and poison bottles, Art Deco bud vases with their heavy, coloured glass, bubbled bases are my trusted favourites, though they come in a huge array of different sizes so I have collected various families of them that are constantly in use.

Collecting a vase assortment is easy, the best treasure will be found at car boot sales, charity shops and even in delicatessens in the form of fancy jars and, of course, vases are very easily found online, too. Here they will almost always come up in a search that uses the much-abused word 'vintage', or just old!

White vinegar and warm, soapy water are used to clean the vases after each arrangement has faded. Old paint brushes are good for feeding into the narrow stems of some of them to ensure that the glass remains clean and free from any algae that forms in clear glass.

OPPOSITE Newly bred yet seemingly old fashioned marbled sweet peas including 'Earl Grey', 'Wiltshire Ripple', 'Nimbus' and 'America'. They can provide a useful contrast to the solid dark and sultry tones.

LEFT Spring-picked violas 'Tiger Eye Red'. Cutting these violas will keep them flowering almost the whole year round. If not dead-headed, they readily self-seed with some interesting results.

RIGHT During a storm all the iris come inside so that their peacock-like crests are not ruined. They are happy on the cool kitchen windowsill and provide a fine view while washing the pots.

Winter decorations

My decorating over the festive season dances a fine line between the fantastical and the plain bonkers! It is an opportunity to let the imagination run wild, a last hurrah before the onset of the properly dank months of winter.

When it comes to Christmas decorations, we have an ever-increasing cast: wire-footed calling birds, glittering geese, marbled baubles, fake (obviously) diamanté-encrusted fruits, costume jewellery butterflies and sea urchin-like creatures, golden moons and stars. Every December out they come, all precious, fabulous, joy-inducing, memory-carrying boxed-up tat! It can easily become an addiction, added to every year from garden centres and stately home gift shops.

There's more than a touch of *Alice in Wonderland*. We have a tortoise, a hedgehog and a flamingo (from New York). Reindeers mingle with childhood Early Learning Centre farm animals – donkeys, Jacob sheep and Norfolk Black turkeys. Sadly, no camels, and so far we have resisted zebras, but Macaroni penguins are another matter; they arrived via the London Zoo gift shop!

This weird and wonderful cast is scattered about, on shelves, peeking out from behind vases, dangling on ribbons and bits of fishing line (a fiddly job that requires a lot of patience and good lighting).

BRANCHES RATHER THAN TREES

Quite recently in a friend's barn, I witnessed the full-scale horror of what it is to have a half-dressed 6m (20ft) tree fall. Luckily it had not yet been dressed in Venetian glass vegetable decorations.

Spruce trees are heavy, dumpy, greasy beings; you are just bringing something into the house to die pitifully. And what happens to the white nets that they are wrapped up in? No doubt many are swirling about in the Atlantic Ocean!

I prefer to decorate the cottage for the festive season with large deciduous branches attached to the beams. These take up no ground space. The best branches to use are mature pieces of cut silver birch, magnolia, twisted hazel and curly willow. The last three can be found at well-stocked florists. I have seen tulip tree branches look incredible, too, with their dried flowers still attached.

To attach branches to a beam or the cornicing, knock in several tacks along their length and wrap a length of florist wire between them. The wire acts as a trellis that the branches can then be wrapped and tied onto.

First place the largest branches that you have in the middle and then add the other smaller ones either side, placing them so that the whole thing looks like one

BELOW Norfolk Black turkeys and Emden goose figures along with many other curiosities, none of whom leave any droppings!

OPPOSITE A crazed *Vicar-of-Dibley-Letitia-Cropley*-styled Christmas pantomime of glass calling song birds amongst a canopy of dried hydrangeas, panicum grass and honesty seed pods festooning a central beam. A good vacuum cleaner to hand is required when all of this is taken down!

large branch with no cut ends on full show; a natural bough of a tree snaking along the ceiling.

You can either leave the branches bare and just add lights and a few hanging decorations to them, or you can build up a canopy of dried alliums, hydrangeas and grass seedheads from them. The latter are so light that they can be draped and pushed through everything else at the end. A mighty mess is caused when it comes to taking all this down, though, so have a vacuum cleaner and dust sheets to hand. You can save the branches and use them for garden staking.

SEEDHEADS

A number of seedheads can be harvested and saved up in early summer especially for this time when their presence in the house as skeletal and ethereal beings will create the boldest of decorative statements.

ALLIUMS The best ones to save in autumn are the heads of *A. cristophii*. This variety is beautiful, large and, unlike others, especially the denser and smaller head of 'Purple Sensation', it will not endlessly shed its seeds.

Spray the heads of the alliums gold, leaving their dried stems on. They can either be arranged in a vase which can be made heavier with some damp sand rather than water, which will rot the stem ends, or they can be pushed into any holes that you have in old beams.

You can also cut the stems to less than 2.5cm (1in) and carefully thread a piece of thin nylon through them with a pin. They can then be hung from the ceiling, like golden stars floating in mid-air – they look especially beautiful en masse, hanging at different heights above a table.

HONESTY It seems to be impossible to buy honesty, so again it needs to be grown, or at least collected from someone else's garden. Just a few of these biennial plants will produce enough seed pods to fill a vase. To grow your own, sow in late spring and it will then flower and go to seed the following year (it's a willing self-seeder on disturbed ground).

The small purple flowers are a favourite of butterflies, then these fade away to reveal little green pods that swell and flatten out like potato chips.

By mid-summer, the honesty pods will be fully developed and beginning to brown. Once they are crispy to the touch, the outer scales of the seed pods can be delicately peeled away. Teasing the bit where the pod's tip meets the plant's stem and gently rubbing will see each pod moult to reveal their ethereal, tracing-paper wings.

Place several stems in a vase on the windowsill where the sunlight can shine through them. They are totally magical.

OVERLEAF

LEFT Damp sand rather than water is used as an anchor for a vase filled with the dried fronds of panicum and Chasmanthium latifolium grass, with the seed heads of opium poppies that have all been sprayed golden. Copper wire fairy lights could be draped through this too, turning it into a miniature Christmas tree of sorts.

RIGHT Honesty seed pods. Each seed pod has been plucked on both sides to reveal tracing-paper silk ovals. A vase of them placed on a windowsill or backlit at night will see the effort rewarded all through the winter.

ABOVE AND RIGHT Pressed
violas create an un-buyable
(possibly for good reason)
découpaged card of
butterfly-like faces. Violas
'Tiger Eye Red', 'Honey Bee'
and 'Frizzle Sizzle Yellow
Blue Swirl'. Lay the freshly
picked flowers onto a sheet
of blotting or sugar paper.
Arrange them so that they
are not touching one another
and then place another sheet
of blotting paper on top.
Sandwich your pressings
into the middle of a heavy
book so that the book's
weight makes them dry
perfectly flat. Within a
month, they can then be
used as card and gift tag
decoration. Glue them
carefully using pva glue.
As well as violas and pansies,
Cosmos also make beautiful
pressed flowers.

HYDRANGEAS People panic about picking hydrangeas and get understandably disappointed when they fail to dry them successfully. It is essential to befriend or bribe someone who has an especially large, mature *Hydrangea macrophylla* to pick some heads from.

These shrubs that often shade out the downstairs window of a terraced house actually thrive on being pruned. When cutting off the flowers, go down to a pair of leaf buds. You will not affect next year's flowers too adversely, unless you choose to totally coppice a bush to the ground!

In fact, cutting some long lengths from it will stimulate a mature plant to send up lots of fresh growth the following spring. If their flowers are not picked then the big mop heads just fade through the winter to resemble a cloud of old burst teabags anyway, so do try to convince someone to let you pick a few!

When to pick is key; too early and they will flop and will not dry. I start picking from mid- to late autumn onward, once the cold nights have begun to turn the flowers to deep purples and maroons.

The cut heads are not hung up but placed in water as single stems around the house and while a few do flop, the majority just dry in the vases over the next few weeks. Once dried they are bunched together and put in a box and more are picked to take their places in the vases, so that by early winter I have several dozen to use.

I once cut a lot of hydrangeas and stored them in a garage for Christmas wreath courses, but, as when left on the plant, the cold meant their colours faded; the warmth of a house will help prevent this.

OPPOSITE *Hydrangea macrophylla* 'Merveille Sanguine', a good small-growing hydrangea for a large pot.

BELOW LEFT Autumn street of front-garden hydrangeas. Once their heads are blush to crimson they can be cut for winter drying. The cutting will help established and large macrophyllas such as these to remain vigorous.

BELOW RIGHT The helichrysums or straw flowers keep their rich colours once they have dried making them very valuable in the house during the months of winter. This one is 'Dragon Fire'.

WREATHS

A late hanging is essential for a keeping a wreath looking really fabulous, so make one as close to when it is needed as time allows. Most people arrive at festivities sober rather than drunk; they will notice if your ivy is limp!

The base of a wreath is the most important thing to get right and here moss is key. Raking yours, or in my case someone else's, lawn is the best way to collect your moss (and a good workout, too). Hand-raked lawn moss is much better than the pricey, stringy sphagnum sort that may well have come from an endangered, far-flung peat bog.

Use a springbok rake and you'll be amazed at how much moss can come out of a lawn, albeit one that has not had feed and weed applied to it. This scarifying job boosts the grass by letting in light and air, so it is a win for both you and the lawn.

You will want at least a good bucket's worth. The moss should be damp, but not soaking wet. Moss, when wired onto the copper ring thickly and tightly, can then hold a huge amount of foliage. The copper ring will last a lifetime, and when you dismantle the wreath later, you can keep it for another time and the old moss can be thrown out in the spring for blue tits and blackbirds to use for their nests.

So, to build the wreath, first tie a wheel of florist wire onto the copper base and attach the wreath's hanger – choose some bright but strong ribbon. Then pat on a generous handful of moss and wire over and under the clump tightly. Having the wreath ring halfway over the edge of a table is helpful for this.

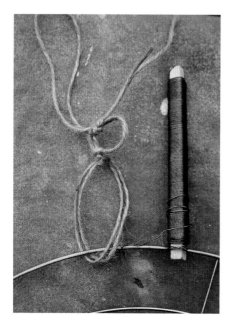

LEFT Before adding any moss attach a hanger to the copper ring. Thick ribbon would be wiser than the doubled-up bit of hessian that I have used here! Also tie the end of the wheel of florist wire to the copper base.

RIGHT Covering the copper ring well with moss is essential for success. It should be tightly bound and thick.

Keep pulling the wire firmly as you go around each clump of moss, so that it is held tight and packed closely. Go over it again and again.

Give it a good shake to be sure that it has all been wired tightly; you want it so that you cannot poke your finger through it.

Once the copper ring is completely covered over with moss, cut off the florist wire and tuck its end into the base.

Now you can begin to add the wreath ingredients. I like to use hydrangeas and fruit. I want it to be full and a bit bonkers.

Red apples, pomegranates and plums, always in clutches of threes or fives dotted around, act like focal points.

I find it easier to get such large and heavy fruits attached first, before adding anything else. Pomegranates get a fake stem. First pierce them with a knitting needle, then shove a small but firm branch halfway up their middles. I cut the end of the branch to a point and then push it through the moss until it comes out the other side, then this is wrapped with florist wire and tied under the moss to the copper ring.

Around the pomegranates shove in some short but strong branches of birch or anything twiggy. These will act like a clawed hand, helpfully cupping the fruit in case any start to dangle downward once the wreath is hung.

With smaller fruits, such as plums and limes, you can use the traditional wired T-junction method. Make sure you buy as firm fruit as possible, then attach these

LEFT Piercing pomegranates with a knitting needle. The hole this creates allows for a wire to be threaded through them so that they can be firmly attached through the moss and onto the copper wire base securely.

RIGHT Stuff the wreath, not just from above, but from the sides too. Look at it from different angles, not just from above, to avoid any gaps.

by poking a piece of thick, green florist wire up the fruit's centre, then another through it horizontally. Then bend the ends that are sticking out on either side of the fruit slowly downward, so that they meet the vertical wire underneath, and twist them all together. Again, this creates a stem that can be shoved into the moss. You might be able to find some crab apple branches with the fruit still attached; these save work, but some will drop off!

Once the big fruits are all in place you can radiate out from them with the foliage. Imagine you are doing floral fascinators that then all join up together.

With the foliage, start off using small bits to densely cover the outside and the inside of the wreath's ring; do not just view it from above, but give attention to the sides, too.

Foliage that looks good and lasts well includes berried ivy, eucalyptus and shiny Portuguese laurel – never spruce! If you can pick these the day before and submerge them all in the kitchen sink for a few hours, it will really help them last for a good couple of weeks.

A few dozen or so branches, ones covered in lichen if you can forage them, and some silver birch can go into the wreath, too. These should be wiggy and messy but not too long, as you just want to create a crest around the dense foliage layer. If I find that I am struggling to get anything into the moss, I make a hole in it first with scissors.

Hydrangeas, with their huge heads, mean that a wreath can be covered quickly, so if you have dried plenty you can almost just use these. They can be stuck into the moss base easily, but it is wise to wire each one to the copper ring as they are so light that the stem ends push out. The dried heads of cardoons look beautiful too and are easily pushed in with their big, woody stems, but if they get wet they will soon fall apart.

Peacock, pheasant and guineafowl feathers can look good, too, used sparingly (I have so far resisted flamingo or scarlet ibis feathers). Again, feathers are best having their ends wrapped with a piece of florist wire in order to attach them to the moss.

Lastly, fronds of panicum grass can make anything that is looking on the dreary side more interesting. Attaching them is difficult, however, as their dried stems snap easily when pushed into the moss, so it is best to add them at the last minute. You can push them in among everything else, so that they cling like fancy cobwebs!

Autumn is arguably the best time in terms of finding material for making huge and spectacular wreaths for celebrations such as Thanksgiving and Halloween before winter sets in. Regardless of the season, the same method of mossing up a copper ring can be used. Live flowers can be added to a wreath as reusable water-holding flower tubes can be pushed into the moss. The peeking tops of these little vessels can then be hidden with foliage. Never use hideous oasis!

RIGHT The completed wreath is hung up on Christmas Eve. Foraged crab apples of the variety 'Evereste' are complemented by smaller hawthorn berries with the faded leaves of hornbeam. Hydrangeas give the wreath a strong heart whilst the seed fronds of panicum grass provide sparkle. Peacock and various bantam cockerel tail feathers are the final finish.

Amaryllis

Amaryllis are fantastic flowers to grow inside for winter clout. They bloom when we are immersed in the depths of darkness, and their flowers can be the focal point of a room. Never plant them as singletons; they look miserable and awkward alone. The more you can grow the better they will look.

While they are an expensive purchase (the best bulbs will cost upward of £10 each, with the more unusual varieties costing more), the largest of these bulbs will each produce three to four flowering stems in succession. These are lean times for ballsy flowers in the garden and what can be bought is usually dire; I don't want white lilies or alstroemeria; amaryllis truly save the day. Their flowers are huge, exotic and clown-like, bordering on the ridiculous, yet marvellous. Amaryllis are the indoor dahlias of winter.

If they are grown in a cool room, each bloom will last for weeks. A pot of several amaryllis bulbs will thus more than pay back the investment. Buy six bulbs and split them into two trios, each of the same variety. Plant the first group at the start of winter and the second a month later, so that you have a succession of flowering.

The pots do not need to be large. Amaryllis actually thrive on being pot bound and having their roots close to the sides, just like agapanthus, but the pot in question must be heavy, as it will have to act as a stable anchor once the bulbs begin to grow and tower upward.

The bulbs don't want to be touching but they can be tightly packed together, like three big goose eggs in a nest. Fill the pot almost to the brim with good potting compost; make sure that is has a hole, vital for drainage, and a saucer, too. The bulbs

OPPOSITE 'Mandela', the deepest red of all amaryllises. Here the four bulbs have also been staked with hazel canes, due to their height.

THIS PAGE LEFT Amaryllis bulbs, their noodle-like roots, despite looking withered, are alive. New roots are often sent out by these bulbs long after they sprout and are showing visible growth.

RIGHT Amaryllis 'Tosca' opening.

often have withered-looking roots, but they're actually very much alive so don't cut them off! Make a hollow to drape them into comfortably as you place the bulbs in the pot, then add more compost, leaving just over half of the bulb poking out of the surface. Firm the bulbs into their pot and then dress around them generously with lawn moss; often by the time the amaryllis are in flower the central heating will have dried out the moss, so it's worth redressing with fresh.

To begin with, amaryllis will want to be kept on the dry side. They take a good few weeks to wake up, but once you see them growing, begin to water them weekly – never water into the heart of the bulb, just sparingly around the base. They are best placed in a cool room on a windowsill; light will encourage their growth.

Interestingly, these bulbs often concentrate their efforts on growing up rather than down; fresh roots are sent down after flowering, so supporting the flowers and leaves using twigs is essential. Twiggy lichen-festooned hawthorn looks beautiful and birch, as always, can be woven around the outside of the whole clump. This will prevent any heart-wrenching snapping. Dead-head the flowers as they go limp, then once all the flowers are over, cut the entire stem off back to the base of the bulb.

While freshly bought bulbs will almost always produce an exciting, huge flower bud, getting bulbs saved from the previous season to do the same seems to be a hard task and one that I have so far not mastered.

Some light was recently shone on this after seeing a mass of amaryllis in a hot house. Pots and pots of them, not flowering, but basking in the heat with their strappy green leaves long. The bulbs apparently need a period of being hot, well watered and well fed with seaweed feed during the spring, after they have flowered. They should then be allowed to dry out in mid-summer. The leaves will wither, the energy will be taken back into the bulb and they will go dormant until they are awoken with steady watering from the start of winter. Keep them snug in their original pots; you could get a teaspoon and spoon out a few centimetres of the old compost each year and add a few handfuls of fresh, mixed with a little potash. A conservatory could provide the necessary warmth, too.

My uncle Chris puts his adored amaryllis bulbs, bought from a Dutch market many years ago, in his dark pantry for the summer with a hessian potato sack over the pot. Then he brings it out again in at the end of autumn to awaken it. The pot is placed in the sink overnight so that it can soak up water and, sure enough, by the end of winter it is in flower again.

However, a lack of indoor space means I cannot really do any of this amaryllis spa therapy, and once they have flowered I soon get sick of seeing them, lingering about the place with their great rubbery green fronds which need dusting. I am ashamed to say I give them away to those with the space to nurture them. But I am trying to be patient and look after a pot that contains three expensive 'Papilio' that look like crazed, flirtatious butterflies when they are in flower.

ABOVE Amaryllis 'Papilio' staked with hawthorn lichen branches.

RIGHT Amaryllis 'Tosca'. This pot contains four bulbs which will give two months' worth of flower presence over the winter season indoors.

SOWING & GROWING

Growing your own plants from seed is by far the most rewarding way to fill a garden with flowers and foliage. Once you have mastered how to breathe life into seed packets, you will be able to diversify into the beautiful, away from the confines of the usual garden-centre-grown bedding plants.

I love selecting packets of seeds, putting combinations together and growing them to cover bulbs for the winter and spring displays, and then combining them with dahlias for the summer and autumn ones. A tapestry is then ensured for both these seasons through using seed-grown annuals, tubers and bulbs.

Seeing the first green, pipping eruptions emerging from a black square of a compost gives an immediate, exciting feeling of being connected to the earth, as it should be. Sowing seeds is very therapeutic, and you will possibly find yourself talking to your growing charges, too!

Bulbs are a total splurge, an addiction of hundreds, in all honesty thousands of them that are ordered each autumn. They arrive in boxes one by one, and soon the space under our kitchen table is full to the brim of this assortment of floral time bombs; it feels like Christmas to open them all up. The best thing about bulbs is that all they need is to be planted and given a pot of good drainage for success – you really can't plant too many of them.

A greenhouse on the windowsill

For those of us without a greenhouse, sowing seeds has to be done within certain constraints. For instance, you will sow things later than those with a greenhouse. It is all about working with the spring season, so that there is a quick turnaround from the windowsill to the outside, resulting in healthy, strong seedlings. The sowing of different seeds, particularly through the spring months, has to be timetabled, to ensure you make the best use of precious window space for your seed trays.

The windowsill will be your greenhouse and often your kitchen table, draining board or doorstep will become your potting bench (with a good dustpan and brush and newspaper sheets for the mess).

If you really want to grow annuals en masse you might want to think about buying a mini greenhouse. It will need to be sited in a fairly sunny place, so it must add to rather than detract from the look of the garden. Slim, elegant glass and steel versions, designed to stand against a wall, look good but will command a high price. Cheaper models made from wood and clear Perspex can be visually pleasing too, but they may not last long. The wooden ones are best, painted with a good wood preserver and put under cover in the winter (most can be unscrewed and collapsed). If you are handy with a drill, make your own greenhouse by upcycling some vintage stained-glass windows and old doors, rubbing them down and painting their frames. This is easier said than done, but they can look beautiful.

I wouldn't bother with one of those plastic, tent-like pop-up greenhouses, though. They are temptingly cheap but in a small space they will not be a visual asset. They rarely last more than a season and tend to blow over in the wind, so if you do get one make sure you weigh down its base with breeze blocks. If you have the space, the half-greenhouse, half-summer house-type sheds with large windows are good. I would certainly have one of these if I could, as it would make propagation much easier.

However, you honestly can grow your garden with just a windowsill (ideally several). It is all that I have ever had. It is another reason why I am not mad about houseplants because if you are using your windowsills as greenhouses there usually isn't the room for these.

Getting growing seedlings off the windowsill and outside quickly is the key to growing healthy plants, but seedlings still need protection from the elements. I've enlisted the help of several miniature cold frames, made from metal and Perspex, bought from Ikea of all places. These fish tank-like squares sit neatly by the door

PAGES 148–9 Sweet Pea seedlings in late winter, happily growing. A miniature greenhouse will provide protection from the elements but allow seedlings the light and air they require to grow properly after germinating inside. One such as this can be easily stored and moved around when in and out of use.

PAGE 150 Use outdoor tables to keep seedlings and unplanted plants off the floor as this will be an aid to them against slugs and snails. Trays and saucers placed under pots of seedlings will help ensure that they do not dry out.

LEFT A seed tray
of healthy cosmos
seedlings that are being
turned around each
day to keep them from
becoming elongated.

and are endlessly in service for seedlings that need to come off the window. I don't
direct sow many seeds as I prefer to have seedlings potted up and growing well, then
I will transplant them into the pots in the garden once they're big enough and the
main pots are ready to be planted up.

Different seeds

ANNUALS

An annual is a plant that germinates, flowers, seeds and dies within one season. They all thrive from having their flowers cut, too, producing more and more blooms until they eventually tire themselves out.

This group of plants is particularly suited to maintaining a good show in the garden. A spring sowing gives flowers from early and midsummer right through to the middle of autumn. And a midsummer sowing ensures that there will be flowers from early to late spring the following year. This sowing will also provide foliage over the winter.

Annuals fall into two groups. Hardy annuals, as their name suggests, cope fairly well with the cold. This means that hardy annuals can be sown both in early spring and in the autumn. Those that are autumn-sown, provided they are robust seedlings by the time winter arrives, and especially if they are given some extra protection – a cloche, for example – will survive the winter and begin to flower profusely from the middle of spring. Hardy annuals that are sown in the spring will flower right through the summer and into the autumn, especially if their flowers are picked.

Half-hardy annuals, on the other hand, are not able to cope with the cold. They hail from warm places. These exotics begin to sulk once winter sets in and eventually they will all be killed by the first hard frost. They do, however, grow incredibly fast, thriving on the longest and warmest days of summer and relishing the morning dews of autumn.

BIENNIALS

Biennial seeds, such as wallflowers, foxgloves and honesty, are so-called because they need longer than one season within which to grow and flower. Some short-lived perennial plants, such as hollyhocks, are classed by some growers as biennials, too, due to flowering at their best in their first season (the summer after being sown the previous autumn).

Biennial seeds need to be sown from the middle of summer to give them time to germinate and grow into small plants. Unlike annuals they won't rush to flower. Instead they go dormant over the winter. Often they will look a bit beaten up, but don't despair of them as they will still be alive. When spring arrives they do not start to flower straight away, but instead continue to grow. Then, as spring starts to sway into early summer, they all begin to come into flower, wallflowers being the

ABOVE Planting up the summer display. Fill a pot almost fully with soil, lay all the plants out on top, then plant down into the pot. There will be enough nutrients for these fast-growing annuals for the first six weeks and then they will need to be fed weekly with a liquid seaweed feed.

RIGHT Wallflowers can look ropey after a hard winter, but come spring the emerging foliage of tulips will hide them while their flowers will peek through, looking resplendent.

earliest, making them the traditional and worthy partner of tulips. Biennials provide huge clout in what can be a dull stage in a garden. It is often easier to buy biennials cheaply in the autumn, as seedlings, rather than sow them yourself.

HERBACEOUS PERENNIALS

Herbaceous perennials include many cottage-garden favourites and meadow wildflowers. These are plants that live for several years and because of this long lifespan, they take their time to grow.

They focus on their root growth in their first year. Some particularly fast-growing perennials, such as echinacea, are classed as first-year flowerers, but all perennials look and flower better with age. I have to have perennials that are good-tempered about being moved about often between different pots, to suit the comings and goings of the more seasonal and shorter-lived annuals.

Perennial seeds can be sown from spring until the middle of summer. Put packets of seed in your freezer for a month, so that when the seeds come out into the warmth the change of temperature breaks their dormancy; delphinium seed in particular reacts to this treatment well. For a small garden, I personally find it more immediately satisfying and, in some cases more cost-effective to buy in perennials as small 9cm (3½in) plants. They grow quickly and will establish well within a summer season. I'd rather focus on growing annuals from seed due to limited space.

NUMBERS

You don't need to grow more than a baker's dozen of each of your chosen annuals to fill a small garden. You may not even need this many if all you really have is a window box.

Remember that you are usually going to combine your seed-grown plants with bulbs, dahlias and perennials, to diversify your pot combinations. It is preferable to grow just a few really strong, healthy seedlings rather than end up being engulfed by hundreds that you don't have the growing space for.

You may be better off ordering seedlings that are tricky and sporadic in their germination, such as thunbergia and violas, as growing plugs rather than growing them from scratch, to save both time and in some cases money. A plug plant usually costs under £1 and these just need to be potted up into square 9cm (3½in) pots and placed on the windowsill to grow on for a few weeks.

OPPOSITE If sown inside in mid-winter and planted out in late spring, the climber *Cobaea scandens* will come into its own by early autumn. Its mutant passion flower-like trumpets reveal its common name, the cup and saucer vine. This is one of the most exotic yet easy of plants to grow from a seed. It will adore the damp and humid days of autumn but the first hard frost will often kill it, unless it is in a sheltered place or moved into a conservatory.

Making the best of a windowsill

We are lucky as our windowsills are fairly deep and generous. In new-build housing, at least in the UK, windowsills are often narrow, in which case you may have to add a shelf along one. Make sure you choose the coolest windowsill in the house.

Line the windowsill's base with a large, water-tight plastic tray that fits it completely. This will maximise the number of seed trays you use and stop water ruining the sill. Water-absorbent capillary matting that can be cut to line the tray is worth buying. The trays and pots will be able to absorb water easily and you can vacuum the matting between sowings to keep it clean and use it for several seasons. You might be talked into buying a heated propagator, but this is not necessary.

Seedlings will always grow toward the light. In a greenhouse, with light coming in from all directions, seedlings grow upward, as they would naturally. On a windowsill this doesn't happen and they grow slant-ward, their young stems elongating unnaturally as they seek the light. So, every day and through the day if you are at home you will have to turn your seed trays around to try to outwit the seedlings' sense of where the light is coming from.

Once germination occurs, you don't want the seedlings inside for very long. For most, three weeks is the maximum before they become useless, leggy and collapsing (central heating makes this worse).

What you want are stout, firm seedlings, like the Two Fat Ladies, Clarissa Dickson Wright and Jennifer Paterson! Incidentally, Clarissa adored cardoons and Jennifer grew pots of herbs on her London fire escape.

To prevent a windowsill start causing your crops to grow gangly, sow seeds, especially the half-hardy annuals, later than the seed packet advises. These annuals are fast growing, but they cannot take cold nights, so I sow most of them mid- to late spring, because by this time the night-time temperatures should be starting to rise and this is what these annuals need to grow well.

A mid- to late spring sowing means that it will be mild enough by the time the small seedlings are ready to go from windowsill to the outside, but they will still need some protection from the elements at first, while they bask in the saviour that is surrounding natural light. A mini greenhouse will protect them from wind and rain, which the seedlings will not have been accustomed to.

You can help the seedlings prepare for this move from the windowsill to the outside by brushing them over with your hand several times a day. This helps their stems toughen up as does having the window open slightly, both during the day and in the evening. Good air flow is vital for healthy seedlings.

An outside table will be useful to put your seedlings on so that they are off the ground; this elevation will help to safeguard them from slugs and snails. I found a wooden potting bench at a car boot sale. It's in use from early spring until late autumn, plonked in the sun in between the larger pots, wherever space allows (in the winter it goes into the dry of my nan's garage). Metal bistro tables will hold fewer seedlings, but they are just as handy and are pretty to look at in a garden all the year round.

LEFT Finely sowing seeds in seed trays will ensure that the young seedlings do not overcrowd one another from the outset, and allows for good air circulation, which is essential for healthy growth.

SOWING CALENDAR FOR A WINDOWSILL GREENHOUSE

Seed-packet instructions don't give much consideration to those without a greenhouse and limited space, so this is my sowing calendar:

SWEET PEAS – sow early to mid-winter; germinate inside, then put into a mini greenhouse or cold frame outside when they reach 5cm (2in). Bring them in at night if temperatures fall below freezing. Sow 30 seeds split between three wigwams, sow one seed into each cell of a root trainer. These will flower in early and midsummer.

COSMOS, FRENCH MARIGOLDS, MILLET AND PANICUM, AMARANTHUS, SUNFLOWERS AND OTHER HALF-HARDY ANNUALS – sow some mid-spring, then at the end of spring. They will catch up, and when sown later they will give you tons of flowers in the autumn, just when you really need them. Sunflowers grow so fast that you can sow them in early summer and they will still flower well.

CORNFLOWERS AND CALENDULA – sow these early to mid-spring for flowers in summer, or if you garden in a sheltered place they will overwinter so you can sow them at the end of summer. Pinch out the growing tips of these summer-sown plants so that they flower the following year; they will do so profusely from spring.

PUMPKINS – I often sow these to replace sweet peas; they thrive on properly warm nights, so sow them in late spring.

OPIUM POPPIES – these are happiest if they germinate where they are sown as they hate root disturbance. Learn to identify them as seedlings – they have smooth, silver leaves that form a loose rosette – then try to work around those that you wish to let grow. Bought seed packets of these should be placed in the freezer for a month to help stimulate germination, then scattered liberally through the garden in mid-spring. Similarly, borage also self-sows but it will be happy to be transplanted

KALE AND CHARD – I sow these as winter pot toppers in midsummer. The autumn nights mean they don't get too big, as they would if sown earlier. This makes them perfect for mixing with wallflowers planted above spring-flowering bulbs.

NASTURTIUMS – the miniature walnut-like seeds of these half-hardy annuals are best sown around the edge of a large pot in early summer. Dob them in about 20cm (8 in) apart. Soaking them the night before will speed up their germination. The seeds emerge with two large elephant ear-like leaves, tumbling over the pot's edge while also bouffanting upwards. Cabbage White caterpillars and black fly aphids are best removed by cutting off the leaf with a cluster, or you can avoid the former issue by sowing them in midsummer. If nasturtiums do get hammered by pests, cutting them back by half will see them regrow impressively by autumn to flower again.

RIGHT Nasturtiums 'Black Velvet' and 'Purple Emperor'. An underestimated cut flower.

SOWING INTO A SEED TRAY, STEP BY STEP

For the packets of seed that you do decide to sow, never sow the whole packet at once. I like to be able to actually count an exact number of larger seeds from the packet to avoid sowing too many and creating waste. A pair of tweezers is helpful for counting them out onto a white sheet of card so you can see them easily.

For seeds that are really tiny, like finely ground pepper, and so impossible to count, just sow pinches with your thumb and forefinger across the surface of a seed tray.

Unsown seed can be saved in an airtight tin for future years, somewhere cool and dark.

Half-sized seed trays, around 22cm (9in) long by 15cm (6in) wide by 8cm (3in) tall, will be perfectly big enough. I avoid sowing into pellets and capsules as I find they dry out very quickly.

Choose trays that are made from heavy-duty plastic. These are more expensive, but you want them to be sturdy and they can be easily washed out between sowings.

Before you fill the trays with compost, you may need to make holes in the bottoms of them, if they have not been premade, to ensure that they can drain. Do this using a screwdriver that has been immersed in a mug of hot water for a few minutes. It will pierce the plastic fairly easily. Make several drainage holes in a zigzag across the whole bottom of the tray.

Fill the tray up to the top with peat-free seed and cutting compost or multipurpose compost. Use a bought compost rather than homemade as this will

LEFT Hardening off seedling cosmos in a mini greenhouse. These seedlings are ready to have their growing tips pinched out so that they bush and produce more flowers as a result. The lid is left open unless a frost or heavy rain is forecast.

RIGHT A cosmos seedling being potted up into a 9cm (2⅓in) pot. Its fast-growing roots will soon fill this.

have been sterilised of any weed seeds. Pick out any large bits to make it finer. If you do use multipurpose, you could always push it through a sieve if it is dry to make it as fine as seed compost, if you have the patience!

Thud the seed tray down a few times once you have filled it to the brim and you will see that the compost level goes down.

Top it up and thud it down again; you want the tray to be properly filled so that the seeds can make the most of the space they have to root into.

Then find an old, smooth-sided book (a Ladybird book works very well) and

LEFT Healthy roots of a cosmos rapidly filling a 9cm (3½in) pot. This plant will be ready for planting in its final pot within a few days' time.

firm the compost with this lightly, moving it into each of the corners. This will make sure that the surface is level and will get rid of any air pockets.

Follow this thudding method if you are sowing into 9cm (3½in) square pots. Larger seeds, like those of pumpkins and sunflowers, prefer a deeper pot for their roots. Place two or four seeds into the corners of these pots.

Most seeds will want soil sprinkled on top of them, as if you were generously sprinkling sugar over a Victoria sponge cake. But some, such as the annual grass *Panicum capillare* 'Sparkling Fountain' and nicotiana, need light to germinate.

I do like to use horticultural-grade vermiculite. It stops seeds hitting the hard crust of the compost's surface as they germinate, helps the seed trays retain moisture and makes the finished sown tray look neat, so I sprinkle this lightly on top of the compost for seeds that can be covered over. If a packet of seeds needs light to germinate then it usually states that they shouldn't be covered, but always read the seed packet's sowing brief. It is by no means necessary to use vermiculite, but I think it does deter those annoying little black sciarid flies that you sometimes get indoors.

As you place the seeds onto the compost, push each one down into it with your finger; the surface will start to look like a carboard egg tray with finger-sized indents. Large seeds like sweet peas, pumpkins and sunflowers should be pushed down to a depth of your finger's first knuckle, but small seeds, such as cosmos and cornflowers, should just touch the compost's surface and be tucked in.

You want to place each seed so that there is a good 2.5cm (1in) around it before its neighbour is placed. What you are trying to do is allow each seed enough space to grow without them crowding one another out. A small pencil is helpful in moving the seeds around if they end up being too close to each other.

By following this rule of generous spacing, you may not have to prick out the seedlings from their seed tray at all between sowing and planting them into their final positions in the garden. Calendulas, millet and cornflowers can easily be grown in seed trays in this fashion.

WATERING

Once you have sown all your seeds, place the tray in the kitchen sink and let it sit in a good 2.5cm (1in) of water. The compost will soak up the water from the bottom of the tray and as it does the seeds will be drawn further down into the compost. Once the surface is visibly moist, take the seed tray out and let it drain on the draining board before placing it onto the windowsill. Most annual seedlings will germinate within two weeks, but give all seeds four weeks before starting to get annoyed with a lack of green activity!

Once shoots appear, turn the seed tray each morning and night to prevent the seedlings stretching toward the light (see page 158).

Continue watering the seeds by placing the tray in the sink. The seedlings will soon tell you if they are drying out, by looking suddenly limp and floppy, but these can also be symptoms of them being too wet, especially if you notice a yellowing of the leaves. Feel the weight of a seed tray or pot to help tell you if watering is needed or not; if it feels light, water, if it feels heavy, wait a few days. Overwatering is, in many respects, worse than underwatering because of the time the soil must then take in ridding itself of excess water. Young seedlings can begin to rot if they are too damp, known as the dreaded damping off.

PRICKING OUT

Some seeds will want to be bulked up rather than be planted straight out into the garden. They need to be transferred from their seed tray into a 9cm (3½in) pot by being pricked out.

Wait until the seedlings have grown a few pairs of large adult leaves before you tease them apart from one another; a teaspoon is helpful for easing them out of the tray. You want to remove them with as much soil around their roots as possible, so water them before disturbing as this will help the roots hold on to the soil. Wiggle a hole in the middle of the compost-filled pot (remember to thud the pot down a few times so that you can fill it up properly with compost), then dangle the seedling into the hollow. Never touch a seedling by its stem as it's far too delicate, instead handle them by their leaves. Firm the compost gently around the seedling with your fingers.

If the seedling has become gangly you can plant it deeper into the pot so that enough of its stem is under the soil to make it stand upright.

LEFT Pricking out a cornflower seedling.

RIGHT Pricking out cosmos. The first adult leaves of this seedling signal that the plant is growing well.

Just like watering the seed tray, place the freshly pricked-out seedlings in the sink filled with about 5cm (a few inches) of water. A 9cm (3½in) pot is used so that the young plant can quickly grow in a new amount of soil that will not hold too much excess moisture. Once the seedling has filled this pot it will be large enough to be planted out into its final pot position, but make sure it has grown a good healthy root ball with lots of root networks showing when you empty it from its pot.

Very fast-growing annuals such as cosmos may need to be potted up into 1-litre (2-pint) pots if room in the garden is not available, but this is one of the reasons for sowing later rather than earlier in the year, as it avoids huge amounts of potting up as you wait for the spring display to properly go over.

PINCHING OUT

Pinching out is removing the growth tip from a young plant. This forces it to put the energy that it would naturally direct into growing upward into growing out sideways. This encourages good bushy growth and more flowers.

A lot of annuals, such as sunflowers, can be transformed by being pinched out as young seedlings. These will then produce over a dozen flowers rather than the expected handful. Sweet peas, too, will thrive on being pinched out as young plants, growing flower-producing side shoots as a result; nip the growth tip using your thumb and forefinger, going down to the pair of leaves beneath it. Cosmos grow at such a rate that pinching out can be done several times over the summer, once they are over 30cm (1ft) high for the taller varieties. Do this especially if they just seem to be putting effort into growing a condensed mass of foliage and are showing no signs of any flower buds.

SAVING SEED

From just one season of growing annual flowers you can save the seedheads for sowing the following year. Calendula, cornflowers, cosmos and opium poppy heads are all worth collecting your own seed from.

Ensure the seedheads are about to scatter their seeds naturally before you collect them up, this is when they have crinkled up and gone brown. Before you store them, lay them out on a tray inside to dry and label envelopes with what each seed is. If you have collected a lot of stems, putting them all upside down in a pillowcase works well; the seedheads will collapse in this and so the seeds will gather in the fabric's corners.

The flowers resulting from these seeds may not look the same as their parents due to the bees cross-pollinating them. As a result, you often will not get the same visual results that the original bought seed packets guaranteed; sunflowers especially like to go back to their native yellow colourations.

SEEDLINGS TO BUY AS PLUGS

These plants either take a long time to germinate or are cheaper and easier to buy in as plugs, don't bother sowing these:

SPRING – *Thunbergia alata*, antirrhinum, calibrachoa.

AUTUMN – wallflowers, fennel, cardoons, primulas and violas, foxgloves.

PLANTING OUT

When planting out, lay all the plants out on top of the filled and prepared container as you want them to be planted. Placing them randomly is better than in lines as you don't want them to look rigid.

I tend to overplant my pots, and I don't think about space too much. The plants, if they are in 9cm (3½in) pots, are then planted a 9cm (3½in) space apart – much closer than they would be in the ground, but this ensures a very full show. Once you are happy with how you have arranged the plants, remove each one from its pot by gently squeezing and tapping it out, and tuck it into the pot's compost. Firm them in using your knuckles to settle them.

LEFT A large tin bath planted with a pair of pumpkins and a row of 'Earth Walker' sunflowers. Both annuals require a rich soil with plenty of manure.

SOWING SWEET PEAS

The day before sowing, I empty my packets of sweet peas into a little dish full of water; the seeds quadruple in size as they soak overnight and look far more promising come the following morning. You don't need to do this, but with soaking, the seedlings should begin to appear after two weeks. The shoots of seeds that are not soaked will take the best part of three weeks to start appearing. I also bring in a bucket of compost from the cold so that the compost warms to room temperature overnight and the seeds feel more cosy when they are sown into it.

LEFT The length and channels of a root trainer encourage the young sweet peas to grow a strong root system over the cold of the winter, which is essential for flowers en masse later in the summer.

RIGHT Healthy sweet peas in their root trainers growing within the protection of a mini greenhouse keeping the worst of the wind and rain off them. These are ready to be pinched out to encourage bushy growth.

Lay some newspaper out on the kitchen table or sideboard and pinch each root trainer together – they have little teeth at either end so that when you fold them up these bite – and place each row in the tray until they are tightly together and ship-shape. Take handful after handful of compost and poke it into each compartment until they are full, then lift and thud them down. The compost level will have dropped substantially in each cell, so continue to fill them with compost and thud them down several times until they are all properly filled and level.

LEFT These seeds are now sprouting after being soaked for several days under a careful watch. Only those that have sprouted are sown. Old sweet pea seed will usually have poor germination.

Brush off the excess compost with a hand brush, then take one seed to sow into each cell, making a hole to the depth of your first finger's knuckle. If you sow a pair, the sweet peas will soon need to be potted on, but just one seed per cell prevents this. This may result in you having some cells that are void of any germination, but reputable, freshly bought seed should ensure a good rate of germination. Cover each seed over with compost. If the compost is dry, water the seeds and place the root trainers on the sink's draining board to drain. After this, put the root trainers on the windowsill in full light.

THE NEED FOR COLD

As soon as more than half of the seeds have started to sprout, it's time to begin making sure they feel cold, because indoor heat will make them become elongated within days, and this has to be prevented as you need the sweet pea seedlings for now to stay nice and short.

The seedlings will be totally fine and strong enough to face the cold outside from being just a week old. However, they cannot face cold combined with wind and hard winter rain, so they need protection from this. The answer is to put the root trainers in a mini greenhouse or cloche that can surround and protect them but still allow them to be in full light. The sweet peas are put outside every day for the cold to nurse them – they will be seemingly dormant for a few weeks but as long as their tips are upright and strong all will be well. Be careful not to overwater them, though, as this will cause them to rot.

Mice can be a real problem when growing sweet peas in places such as greenhouses. In these situations the top half of an old hamster or bird cage is a useful protection over the sown root trainers. Alternatively, sowing them in the house will hopefully mean this isn't an issue.

Once the seedlings have grown four pairs of leaves each, their growing tips can be pinched out one by one. Again, this action prevents forming seedlings that are tall!

The seedlings should look strong and be of a good Bird's Eye garden-pea-bag green. If you take the growing sweet peas from their holder in their root trainers you should by early spring begin to see the ends of the sweet pea roots growing through the cells' bottoms. These can be pinched off, too, a process called air pruning. The root trainer cells with their ribbed sides will be encouraging the little roots to grow vitally downward and the roots showing are a telling sign that the seedlings are bulking up well. You can begin to feed the sweet peas with a seaweed feed while they are in the root trainers if any begin to look off colour, but by early spring it will be time to try to plant them.

BELOW Sweet Pea 'Blue Velvet', a recently bred Spencer type sweet pea. Like all of this group, it has a longer stem, making it especially good for cutting.

UNDERSTANDING DAHLIAS

The most important thing to learn about growing dahlias is how to ensure that their tubers do not rot at any point of the year. Imagine you are dealing with a sack of firm potatoes that you must not allow to become rotted mash; almost always the cause of this happening is too much water. The tubers of dahlias can cope with a surprising degree of cold, but they cannot survive being in waterlogged conditions – ever. This all hails back to them being plants of Mexico where the soils can get very cold but they are usually very dry. This is what made the dahlia evolve into growing a tuber in the first place; it acts as a big store of water for the growing plant, prolonged excess wet sees its outer skin unable to cope and it will fatally rot.

SLEEPING DAHLIAS – WINTER

Thanks to milder winters, there is now a more relaxed attitude to leaving dahlias in the ground all year in most parts of the country. If mulched over with compost like a molehill duvet, a tuber can survive without it having to be lifted in the traditional manner. The top foliage of the dahlia will be completely killed but the tuber will remain alive underground, ready to sprout again once spring returns.

Pot gardeners, however, out of practicality, must lift their dahlias rather than leisurely leave them in for the winter. If your pots are in a sheltered garden, of course you could mulch them but these pots can't then be properly planted with bulbs for the spring display, as there isn't enough space for the spring bulbs and dahlias to co-exist. So, for this reason, all my dahlias are lifted to allow for the pots to be completely replanted for spring.

LEFT Digging out tubers from the tin baths in the autumn. If the tubers are wet then they must be taken inside to dry before being stored for the winter in boxes somewhere dry to ensure that they stay firm and healthy.

RIGHT Awakening dahlias in shoeboxes. When stored in cool and dry rooms indoors dahlias will overwinter well. These are already starting to bud as they sense spring's approach.

The dahlias start to be cut down as soon as they begin to look tired and pot by pot they are lifted. The vital reason for this, rather than waiting, is so that the wallflowers and kale can get their roots down into the pots properly before the weather turns. It is also a nicer job to lift the dahlias and cut their foliage down before the frost has blackened it and turned it to a smelly slime.

Because I have not watered the pots for a week or so, any soil attached to the tubers readily crumbles off and the tubers soon look like smart, freshly bought ones. If the weather is wet, as it often is in the autumn, you need to ensure that the tubers are dried off fully before being packed away for winter. The easiest way to do this is to lay them all out on an old sheet, under the kitchen table, for the best part of a week.

The tubers can then be packed into sturdy shoeboxes, with each dry tuber wrapped in newspaper. Place these boxes somewhere that is cold but not damp, such as a spare room or in the pantry, but do not put them in a cellar or a damp outhouse as such places will encourage rotting.

But what about labels, you may be thinking? Go through each plant before you do any cutting down and tie a luggage label written in pencil of each dahlia's variety. Tie this onto the thickest visible stem at its base, that way if you get carried away and hack them all down in a frenzy, the labels will be there to identify. I have stopped labelling my dahlias because in time you learn the individual varieties of your favourite ones through recognising their different leaves and I now only grow the trusted varieties that I like. Any new ones that I trial I will grow in separate pots, coal bucket-size that hold about 4 litres (7 pints) of compost, and if I like them then their tuber gets saved and chucked in with the rest in a box to keep over the winter, to appear somewhere the following year.

AWAKENING THE DAHLIAS – SPRING

I probably grow, as a minimum, 30 dahlias each spring, which might sound like a lot but that is what comes with being addicted to certain plants. I'd grow hundreds of dahlias if I had the space and the time!

The dahlias go into their summer pot positions not as dormant tubers but as well-grown plants. The tubers that have been stored in the shoeboxes are reawakened in mid-spring, no earlier because they grow very quickly and yet they cannot be put outside properly until late spring, when all risk of a late frost should normally be gone.

I used to pot each dahlia tuber up on its own, into a square plastic 2-litre (4-pint) pot filled up to more than halfway, then the tuber would be placed in and covered around with more compost. Never plant a tuber deeply, as they like to be just under the compost's surface. This style of starting the dahlias in growth was fine when I did not grow many, as half a dozen such pots could be put onto a windowsill to sprout and grow.

BELOW A 3-litre (5-pint) pot-grown 'Bishop of Auckland' dahlia, ready for planting out. Not overwatering is key for healthy growth.

ABOVE Starting dahlias in pots will help protect them from slugs and snails, but their foliage may still be at risk when they are planted out.

LEFT Dahlias in crates ready for planting. Growing them in these containers means that both a large number of plants can be grown and that they can be easily put outside and taken in during the weeks of hardening off in early spring.

TRAYBAKE DAHLIA CRATES

Now that I grow more dahlias, however, I use crates to start them all off in; the plastic sort to be found in the fruit and veg aisle at the supermarket are ideal, wooden ones are liable to break once they are full of compost.

Obviously just one crate would take up a whole windowsill, so the dahlia growing happens in one of Min's spare rooms upstairs. The crates are lined up on old carpet-covered sturdy boxes to meet the height of the window so that the light can properly get at them. If I did not have this option, I would have to seek out a greenhouse locally to adopt or just go back to growing fewer dahlias!

The crate is lined with old plastic compost bags so that the compost does not drop out of the slits. The crate is then filled to 2.5cm (1 in) below its top and eight dahlia tubers are nestled down into the compost. The compost wants to be quite rich but light, it could be a 50/50 mix of new and homemade, if you have that.

Pointed scissors are daggered into the crate's base once they have been filled with compost bags; this piercing ensures essential drainage.

If the compost you are using is moist to the touch, then you do not need to water the dahlias until they have started to sprout and grow.

The compost as it is will be enough to stir the tuber into life, but if the compost is dry to the touch then you should water them – about a mug's worth per tuber, excess damp will easily make dahlia tubers rot at this stage, too!

By late spring, each crate will be starting to have a good haze of green within it. On fine days they can be taken outside as the fresh air will do the growing plants good; if the forecast is for a warm night you can leave them out. Some tubers will sprout dozens and dozens of stems; thin out any that look weedy to give the larger stems extra room to breathe. Pinching out dahlias is important to create very bushy plants that will, as a result, flower more. Once each stem has four healthy pairs of leaves, pinch the growing tip from all of them – this is less important for the decorative types, I have found, but essential for a number of the singles and anemones, as it really makes them much stronger plants. If you plant your dahlias in early spring you may have to do this more than once.

When the time comes to plant the dahlias into the garden's main pots, treat the crate as if it is a traybake and use a trowel to divide each growing dahlia from its neighbour into its own slice of dahlia and compost, taking the trowel to the bottom of the crate and gently lifting the dahlia up and out. Dahlias seem to have a transpiring reaction to hot weather that makes them look like they really need to be watered, looking all floppy and overcome but as soon as the sun goes in, they will perk up again.

I only water the dahlias routinely once flowering starts, as this signals they are establishing well. For the first month of being planted in the dolly tubs watering happens less than once a week; remember the need to ensure that the tuber is not overcome with too much moisture at all times! However, usually by the middle of summer growth is at full pelt and in the case of extreme heat, watering can be as much as twice a week for dahlias in pots with an added liquid seaweed feed chucked in each time. Seaweed I have found to be the best feed for dahlias; in the dahlia-growing fields in Holland whole crops are sprayed with it.

BELOW The dahlias are removed from the crates as if they are slices of dahlia cake! Some, despite having a large amount of top growth, may not have grown much of a root ball. Don't be too alarmed, as this seems to be a growth habit that differs between varieties. Always plant your dahlias out on cool days or in the evening though, so that they can settle without drooping in the sun.

Endless bulbs

The dahlias seem to thrive on awakening to a facemask of autumn dew on their petals and this is the time of year that sees fronds and leaves needing to be pinned back with any sprigs of birch that are left from the almost-depleted bundles so that you are not brushed with wet when you leave the house in the morning. The pots brim over; it is a beautiful look of totally dense chaos offering endless dahlias to cut and sparklers of seeding grasses.

Depending on how you choose to implement your spring display, autumn can be

LEFT Eremurus. These high-end empresses of bulbs (with a price tag to match) need as much grit as you can give them to thrive. Here fresh bulbs have been placed with their roots overlapping each other, just as they would do in a natural clump. Doing this allows a pot to have several exciting and glorious flower spikes.

very relaxing. If all you want to do is to plant tulips, the entire garden can be allowed to fade away until the first hard frost, then you can lift the tubers of the dahlias to store and plant your tulip bulbs as late as early winter. This is because tulip bulbs do not need to start growing their roots until late winter; indeed, such a late planting of tulip bulbs will safeguard them against fungal spores that will be killed by the colder temperatures of winter frosts.

If, however, you want a spring with more floral diversity, the summer show cannot be allowed to sit and fade out into late autumn because, unlike tulips, other spring bulbs and plants (the pot toppers) will want to start putting their roots down before the winter chills the soil.

Because of this, you have to begin to execute the summer pot show and plant the spring show by mid-autumn for wallflowers and kales to be happy. It's a hard thing to do, but it's essential if you want a lavish look next spring.

The most vital thing for the bulbs is for their pots to have good drainage; every pot needs to be checked to ensure that the drainage holes are free from any blockages before they are planted up.

Freshly bought bulbs have all the energy within them to flower well; the only thing that I add to the pots' compost at this time of year is some blood, fish and bone meal – a few handfuls well mixed in – as a slow-release fertiliser.

ABOVE New and old tulip bulbs. The old bulbs (right) are unlikely to flower well again as they have spent energy on growing baby bulblets that won't flower for around two years, so larger and freshly bought tulip bulbs are required each year to ensure a full and exuberant display that would otherwise be one largely of leaf!

Bulb lasagne

Bulbs allow those with small spaces and especially those who just have pots to really go to town with flower numbers. You can literally pack dozens and dozens of different bulbs into large pots and the result will be a spring display that goes on for months, with minimal attention compared to the summer show. Bulbs arrive in boxes from early autumn onward, so store them indoors somewhere cool, away from any radiators. It is vital to order your very favourite varieties of bulbs early as they will often sell out, tempting as it is to wait for them all to go on sale!

People who haven't done a layering of several bulbs in one pot before, known as a bulb lasagne, can find this concept often quite odd but once you see the abundant and successive effect it results in, you'll be hooked on planting bulbs in this fashion each autumn.

It makes complete sense to layer bulbs in a pot and the bulbs will not mind being under or above others either; as each one sprouts it will just grow around any bulb above that it hits on its way to the pot's surface. It is a pattern of compost then a layer of bulbs, then compost, then a layer of bulbs.

Most bulb lasagnes consist of three layers of bulbs with handfuls of compost that create gaps that are about 10cm (4in) in between the layers.

In tall pots such as dolly tubs or terracotta long toms, the pot should be more than halfway filled with compost before the first layer of tulip bulbs is positioned; remember that no bulb should be touching its neighbour but I do still pack bulbs into the pots in an almost cheek by jowl fashion. In shallower pots, such as tin baths, the tulip bulbs can go in as soon as the compost has covered the pots' drainage material by about 5cm (a few inches). Pots such as these can usually only manage two layers of bulbs simply due to their lack of depth.

OPPOSITE Looking down at the garden in autumn. All the pots have been planted with winter and spring pot toppers with spring bulbs under them by now. The wisteria looks resplendent in its golden tones, befitting an ancient Roman's leaf crown.

ABOVE Depending on the varieties being chosen, one layer of tulip bulbs per pot may often be enough, especially if the pot, once filled, is going have wallflowers added.

USING AND MANAGING BULBS WITHIN A BULB LASAGNE

Suggested numbers are based on pots the size of a dolly tub, dustbin or terracotta long tom.

CROCUS – flowers in early spring. Dead-head as they fade, but leave the foliage as these leaf fronds cascade over the sides of the pot. These are planted as the final layer to a bulb lasagne, the one nearest the pot's surface, 10cm (4in) deep. Plant 30 to 50 bulbs per pot. Lift when the bulb show is over, with their foliage still attached, then dry hung up in an onion bag. The same method of preserving bulbs from one year to the next also applies to the iris and hyacinths below.

IRIS RETICULATA – flowers in early spring. Dead-head as their flowers fade, letting the leaves grow tall and spear up through the tulips. These are planted as the final layer to a bulb lasagne, the one nearest the pot's surface, 10cm (4in) deep. They can be planted in the same layer as the crocus bulbs. Plant 40 bulbs per pot. Also plant as trios of bulbs in small terracotta pots for an auricula-type display to bring inside.

HYACINTH – flowers in mid-spring. Dead-head the entire stem of each bulb once the flowers fade, cutting off all but the leaves. The bulbs are often pushed up by the emerging tulips that are in the final layer of the bulb lasagne beneath them. The tulips will benefit from having the hyacinths bulbs removed as they begin to surface so that there is more room. Remove the bulbs with their foliage attached and save for next year. These are planted as the middle layer of a bulb lasagne, at about 20cm (8in) deep. Plant five or seven bulbs per pot. Also plant in small pots to bring into the house so that their fragrance can be enjoyed.

TULIPS – early tulips flower from mid-spring, late tulips from late spring, but as springs are getting warmer, late and early varieties are clashing more and more. Plant two layers of tulips unless you plan to over-plant with small plants such as violas and primulas. For pots that include these, just one layer is sufficient; two will mean that the plant toppers quickly become suffocated as the tulips emerge. Tulips are planted in the second (middle) and the third, final layer of a bulb lasagne – the middle being at 20cm (8in) and deepest layer being 30cm (12in).

You can save your tulips if you want, but for any of them to have a hope of flowering next year, snap off the seedheads as the petals drop and lift with their foliage still attached. Put them all in a loose-stringed onion bag and hang up, away from mice, somewhere light, airy and dry for the leaves to frazzle. You can then sort and keep the largest bulbs in the summer, any that are still 2mm ($\frac{1}{16}$in) thick upward in belly and feel firm should flower again, but those that are smaller are not worth planting so discard these; they'll flower again, but not for two or three years.

I tend to plant new tulips every year because the warmth of a pot encourages the big tulip bulbs you have planted to make lots of baby bulblets; the bulb's energy goes on these often at the expense of next year's flowers. I offer the finished bulbs to friends with larger gardens or allotments where they can be planted to either flower again – or not – at their leisure!

RIGHT A tapestry of iris 'Purple Hill', crocus 'Orange Monarch' and fennel in mid-spring.

ALLIUMS – posh onions! The stalwart *Allium* 'Purple Sensation' flowers with the late tulips from mid-spring and then the later-flowering and larger alliums, most notably *A. cristophii*, will help to fill the late spring–early summer flower gap. If you are careful at removing the other spring bulbs and delicate at planting out the dahlias and annuals, then the alliums can continue to give presence to a pot as the summer show puts down its roots. The purple flowers fade to leave shimmering green seed pods and by summer these sparklers will have gone brown, making the garden feel autumnal and tired, so tug them up by the stem. They usually break easily from the bulb that can be left in the pot until the autumn. Take these sculpture-like seedhead creations inside, somewhere warm and dry, as they look beautiful as natural decorations over winter.

Alliums will begin to send up their strappy leaves in early spring. By the time the flowers start to open, these will look very tatty. It is perfectly ok to cut them off either halfway or completely by this point as their leaves will have given the bulbs enough energy by now for next year's flowers to be ensured.

Allium bulbs quickly multiply, which is why they are cheap to buy. When you do lift them, discard or give away the small bulblets that will have grown around the original bulb in the middle and keep the largest, as these are the ones that will flower again next spring – the smaller ones will just send up leaves. Plant randomly within the layers of tulip bulbs. Because of their large leaves, just five will be more than sufficient for a mixed bulb lasagne.

FRITILLARY – *Fritillaria meleagris* is a true fairy lantern, hanging off a slender stem with its purple guineafowl-patterned plumed bell. To see them at large in unmown grass, where they look at their best, admittedly makes me crave a lawn, however, they will only be perennial on hard but damp soils where they will multiply if undisturbed and allowed to go to seed. They rarely prove to be perennial when they are grown in pots, even if you save the bulbs with their foliage left on. Surprisingly though, these bulbs looking like bits of popcorn are cheap. They will sometimes attract lily beetles, so be on the watch for these. They are so exquisite that it is nice to plant them in several small terracotta pots that can be brought into the house each spring; they make good cut flowers, surprisingly, if kept in a cool room. Plant 20 in large pots in the final, top layer of a bulb lasagne, mixing them with crocus and iris.

The larger and much more expensive imperial fritillaries are, alas, to be avoided in small gardens as while they may be incredible to look at, most smell strongly of fox piss!

OPPOSITE Bumblebees on *Allium christophii*, whose common name is star of Persia. Here all the tatty foliage has been removed but this will not affect next year's flowers.

LEFT Tin baths that have been freshly planted with tulip bulbs. Chicken wire and chilli flakes are often required as a necessary deterrent against grey squirrels, who would eat the lot over a winter weekend given the chance! Pots planted with bulbs may need to be watered lightly from early spring onward if there is little rain during a dry winter and early spring.

SQUIRREL PROTECTION FOR BULBS

Grey squirrels are a plague and see pots filled with tulip bulbs as a Michelin-star restaurant, relishing the easily scratchable soil that allows them to carry away bulbs with ease. The best protection is placing thick wire netting over the pot and sticking it down with the longest tent pegs you can find. Squirrels can easily chew through cheap chicken wire, but rusted fireguards and even old and helpfully circular barbecue grills, will resist squirrel chewing, and, being rusted, they don't look too bad as pot defences either, god forbid anything new and shiny!

For pots that cannot be covered flat with wire, such as those with pot toppers, you can make cloches using silver birch and hazel bent over and stuck into each opposite side of the pot, then plaited to create a beehive shape. This you can then cover with

LEFT A large oval,
galvanised bath with
the striking presence
of silver cardoon leaves
that have stood the
test of winter. They
are now joined by
hundreds of tulips,
like sprouting beaks.

stronger florist chicken wire that is available painted in green or black, helping to visually enhance such a creation. Ready-made wire cloches are available and will save a lot of faffing about and they look ornamental, too. Again, stick their bases down with tent pegs; a squirrel will try hard to get at your bulbs once it knows they are there.

Chilli flakes bought in large bags are a defence that will work once a squirrel begins to dig into a pot and they can be easily sprinkled around the pot toppers, but I prefer wire protection mainly because the chilli flakes need to be applied very generously to be effective. They also resemble vomit on the pot's surface, which isn't a nice thing to look at by the door, oh dear.

POT TOPPERS

Pot toppers for winter foliage and spring flowers are planted once the bulbs have been planted to give interest through the winter months and to complement the bulbs when they flower. The latest that they can be planted is mid-autumn, because they need to root themselves into the pots' compost before it gets too cold. Each plant needs to be firmed down around its base with your hands because a hard frost can lift a root ball as it expands the soil particles.

VIOLAS – either order as small plugs in late summer and pot each one up in a 9cm (3½in) pot to grow them on or sow them from seed earlier in the summer. The first option is easier as their germination can be sporadic. Plant them around the edge as, being low-growing, they can easily be swamped by bulb foliage, or plant in separate medium-sized terracotta pots en masse, though the more root room they have, the more flowers they will produce. Dead-heading or picking them for the vase is essential to keep them in flower; this also stops them from becoming leggy. If you do this they will flower for almost a year.

WALLFLOWERS – these are the trickiest of the lot because they sulk easily and then they look tatty, but persevere with them even if they look awful because as soon as spring properly takes hold the flowers are the perfect, scented backer to tulips. You will have to groom them over the winter, removing yellowing leaves. I have found personally that they don't like to have their tips pinched out. If you can find them for sale as plug plants with well-grown roots this is a better option for pots than buying them as bare roots, which need to be planted straight away, ideally in late summer, when your pots are likely still to be summer party mode. Bare-root bunches are good to shove into a bed at the allotment or larger garden, however, and they are very cheap in comparison, too. Plant five good-sized wallflowers per pot. These may require a cloche to protect them from woodpigeons and any resident chickens!

BRONZE FENNEL – cheaply bought as small plants at any good garden centre or farm shop, fennel gives a unique feathery presence through the winter and into the spring; they are also perennial. In summer, cut each plant back, then pot up and replant with cosmos seedlings or something similarly airy, to give foliage presence to the garden over the summer. If you are worried that it will self-seed itself, cut off the flower heads once they have faded from being yellow. Plant one fennel per pot.

KALE 'REDBOR' – like a crinkled seaweed that never fails to get complimented on. Sown much later than it would be for a vegetable garden, in the middle of summer. These little seedlings grow fast, forming an attractive but bonsaied rosette of crimson leaves, due to them growing late in the season. Each seedling is potted up into its own 9cm (3½in) pot and once they start to get going, they are fed with a seaweed feed when I water them so that they do not become hungry; wallflowers

TOP Polyanthus varieties with tall stems make elegant additions to spring pots and may flower through winter in a sheltered place. Cutting them for the vase will make them flower well for months. Those of the F1 Stella series, such as this deep pink sport of 'Stella Neon Violet', hail back to their grassland primrose ancestors.

ABOVE Kale 'Redbor' in flower.

RIGHT The healthy leaves of kale 'Redbor' are ensured by feeding them a liquid seaweed.

SOWING & GROWING

will also appreciate a feed as they too are members of the brassica family. The kales stand up well against the winter, looking especially beautiful in the snow and frost. Once spring dawns, the lengthening days encourage the kale to send up delicate rods of blossom; pinch these off and they will send up more. The blossom is a lovely light lemon-curd yellow and the bees adore it. When flowering has finished, the little plants quickly succumb to aphids, so remove them. It is important when growing kales in pots that you take a year off having them now and again to prevent gross root maggot fly, whose lava will eat the kales' stems and ruin every single plant.

CARDOONS – the most architectural and arching of frond-like leaves, elephant grey, commanding, beautiful and useful in a large pot where they give William Morris-leaf-styled presence throughout winter. In spring they complement parrot tulips, especially, matching their scale and arching stems. You can find small seedlings for sale that are small enough at 1-litre (2-pint) pot stage to not be too big for a pot to handle. Cardoons are versatile so you can constantly lift and re-plant them. If the leaves get too big for the surrounding bulbs or plants, just cut them off at the base and more will reshoot. If you do not cut them back they will swiftly become giants. After the spring show has finished, they can be lifted and potted up into 2-litre (4-pint) pots or given their own coal buckets to see out the summer in, then they can be replanted again in dolly tubs come the autumn. Often in the summer they suffer from blackfly but any affected leaves can be cut off entirely. If they can be allowed to become large and bloom, the spikey goblet flowers are a feast for bees in late summer.

OPPOSITE Cardoon foliage giving presence through the winter and spring magnificently. Cut the leaves for the vase to stop these plants becoming giants.

BELOW Bee cardoon dive.

Staking and support

The gathering of birch takes me to the dark and neglected places in our town. It sees me become a strange, ant-like, hypnotic creature for several days in mid-winter. I hope it entertains some people who see me carrying the broomstick bundles under my arm along the roadside, which I gather until I have around 100 stems, enough to stake all of the pots for the coming summer season and to make wigwams and amaryllis nests.

Birch is my best garden prop; it supports and structures the garden. Without it, the summer show would be a collapsed, tipsy affair. Very few of the dahlias, cosmos, gladioli or sunflowers would survive without these halos of support. As they get taller and they begin to flower, their stems lean and they often either snap or topple over. The stormy winds are ruthless to a flower garden, one of its worst enemies, and it will be a demon to any fragile or heavy growth that has been left unguarded.

Birch is a nurturing claw to many flowers and stems, the first flowers it supports of the year are the heaviest tulips, the parrots. It also creates vitally tall vertices in the form of wigwams for sweet peas, being generous in its twiggy nature for tendrils and tips to curl up around with ease.

The thing about garden staking is that you do not want it to put the garden in a straitjacket, it needs to still allow plants to billow and seem free from restraint. The material also needs to be from trees that won't attempt to root and grow once their

ABOVE Hens on birch. It is essential for the birch to be picked before its sap starts to rise. Gather more than you think you need and stack it in a corner, in tied up bundles. It won't get brittle until summer.

stem ends have been pushed into the soil. I avoid using willow as this roots readily, dogwood stems often do too. Once they root, they will begin to take away the pot's nutrients and this will then starve the plants that these supposed dead aids should be supporting. Willow, while often seen as the go-to branch for weaving, I do not find to be easy to work with in the garden as it often snaps when you try to twist it into a shape. However, if you can only find willow, you can prevent it rooting by stripping the bark away from the stem end that you are going to place into the soil.

Often staking needs to be put in place before it is needed, ideally so that the plants grow though it and then hide it with their leaves, but if you can make staking the garden a beautiful art in itself you will enjoy putting these supports in place. In particular I have found birch to be hugely versatile and ornamental.

The birch brush wood are actually self-sown saplings of silver birch, but being wild infants of these silvery trees they are instead of a muddy chestnut brown and they grow in urban areas, often in abundance on scrub and waste land in particular. It tends to spring up anywhere where the ground has been disturbed during the past few years and it's the young saplings that you want to coppice, as these are different to branches from mature birch trees. The young trees grow a single upright stem that is very well branched and happily pliable; it can be twisted and plaited but the stems hold a useful tension.

ABOVE Birch sticks do not always need to be plaited; they can be used in a simple pea stick-like fashion, with them lined around the inner rim of the pot but also zigzagging through its middle.

OVERLEAF The garden at the in-between stage of autumn going into winter. The tin baths have been planted with bulbs because the wallflowers, violas and kales freshly planted above them need these last weeks of autumn to put down their roots while it is still mild.

HARVESTING GARDEN STAKES

Coppicing is the way to cut these young trees. It will not kill them, indeed, it will encourage them to grow a bushy framework rather than a tall one. Coppicing is an ancient way of woodland management that once saw doormice and wildflowers flourish, as it opened the canopy, letting more light get to the ground.

Hazel also makes for good garden staking and this will thrive on being coppiced often, as this keeps these trees vigorous it encourages them to send up fresh rods. You can cut hazel and birch stems down to 12cm (5in) off the ground and they will soon re-sprout. The cutting of them has to be done in early winter, because with springs being so variable now the tree sap often rises earlier, and if you cut them in early spring you risk what you have harvested coming into leaf and you'll then have to strip off the buds!

You don't need to harvest any stems that are more than 7.5cm (3in) wide; you want most of the stems to be as thick as a Sharpie pen, so this task can be done with a good pair of secateurs. Take some hessian string to tie up the bundle tightly so that it can be easily staked up at home. It will remain pliable until early summer.

Gathering birch is a task that needs to be undertaken each winter for fresh supplies as old supports look tired and become brittle. It makes for good kindling when you are taking the garden down for winter.

If you live somewhere rural you may be able to find farmers and shooting estates who sell bundles of birch and hazel brushwood for staking, as it has become more popular to use in recent years. Some of the good garden centres may sell it seasonally but ask them well in advance to be sure.

If you are like me and you have to forage, then legally you need to always try to seek permission from the land owner that you are foraging from – a hard thing to do if you are passing land that is seemingly abandoned and covered in rubbish and rubble, but use your common sense.

The first thing to do is to get in touch with the council's parks department, who should put you in touch with the local wardens of parks and public land; local wildlife trusts who own nature reserves and may be managing habitats in which birch is invasive can also be very helpful and will be glad of you doing a morning's volunteering of birch cutting. If the birch is growing at the back of a supermarket, be bold and ask the security guards or the manager for direct permission, explain what you want to do with it and tell them you'll bring them some sweet peas come the summer as a thank you. They will think you are crazy but if they have some humour and heart then they'll just say yes, feel free, of course. If not, leave it. If the answer is no you will be deemed mentally insane and they'll probably escort you off the premises and you won't want to show your face there for several months afterwards, but at least you were polite and asked in the first place!

OPPOSITE Me, looking like a *Wicker Man* extra, but alas no Britt Ekland is in sight! Foraging birch in the winter for plant staking and wigwams.

ABOVE Beyond the brink and far from the madding crowd, carrying birch. Hessian string is essential for helping to gather the birch into carriable bundles!

HOW TO MAKE A BIRCH HALO

Birch halos are used in the dolly tubs and dustbins for staking dahlias and annuals. They are usually made once the dahlias have been planted.

Take 15 rods of good twiggy birch that are about 45cm (18in) long, then push each into the pot around its inner rim until they have encircled the whole pot.

Take an upright branch and curve it so that it is horizontal, just above the growing dahlias, then take another and twist these together. Do the same to each, plaiting them until you reach where you started, then thread the last stem ends back into the created halo. The more practice you have at this the better you will get at it.

You don't have to always plait the birch, you can simply insert and thread it all the way through a pot of dahlias in zigzags, but you'll need to do this quite densely using bits that particularly look like gnarled outstretched claws!

If I could not forage any natural supports I would use bamboo, but I would not have it as it is all sandy yellow and shiny new, nor would I allow it to age into the awful shades of old smokers' teeth. I would paint these canes with outdoor wood paints in colours – perhaps indigo blues, greens, perhaps even oranges! Then I would put five or seven in each pot and spider-wed taut hessian string from each cane to create support for the leaves and stems to grow up through.

If money is not an issue, those handsome cast-iron supports that rust beautifully take all this faff away, but I have yet to ever buy any. There are some lovely blacksmith-made creations out there, though, that will last a lifetime and beyond. I know several gardeners who use small and huge lobster pots (I love the idea of turning a large one into a broody hen coop), spheres and half-moon-like curves on legs en masse. Given the amount of staking required here, such an investment may well be worth it if you are short on foraging time but I do love a birch nest and the crackers effort that is needed sometimes to get it!

OPPOSITE Woven wigwams in crazed weaver bird fashion. By the door is the biennial velvet pin cushion-like Sweet William 'Sooty' growing in coal buckets.

BELOW RIGHT Dahlias freshly planted with silver birch unwoven and stuck into the pot around its inner rim. The cardoons have been carefully left while the tulips and the other spring bulbs have been taken out.

BELOW LEFT Upright sticks of birch have been layered onto each other and plaited together, creating a protective outer halo to the plants against summer storms. The dahlias will outgrow it but it will still provide vital support.

WIGWAM

The empty unplanted bins or dolly tubs should be enriched with fresh manure and good compost – chuck in several good handfuls. Then thoughts must be turned to the vital tower of support that the sweet peas will climb up, the wonderous wigwam!

There is something very beautiful and simple about what the presence of a stick wigwam brings to a garden, even in a sleeping winter garden. You put a wigwam up and you go oh thank god for that, a bit of proper action going on in this godforsaken hole, let's do another! Instant elevation.

My preference, as with everything staking-wise in the garden, is birch, which is seemingly the best branch for a sweet pea as it is very twiggy and of a good, rough texture. The tendrils sense this and within a day you will see that they have wrapped around the little branches like baby octopus tendrils.

Compare this to the unkind and slippery bamboo cane, where you see sweet peas drowning because the canes just do not have a climbable surface; they try to engulf them but it is not a willing support and so the stems collapse at the bottom and grow into each other. Far better if you are without birch or brush wood is to do what they do at Easton Walled Gardens in Lincolnshire each spring and grow sweet peas up a circle of upright stock netting, which seems to work very well for this acclaimed sweet pea garden.

If all you have are canes, then you will have to tie taut hessian like a spider's web around and around them to create something in between each cane for the plants to grow up.

The nice thing about using a large pot for a wigwam is that you cannot fail when it comes to putting the branches or canes in a perfect circle shape, I shove in five very tall stems of birch that are of a good 3m (9ft) with 15cm (6in) going into the pot firmly. They are tied together at the top – and for the record you are now having to stand on a chair to do this.

The birch wigwams look nice at this stage, just natural, like upside-down witches' broomsticks or crazy unfinished weaver-bird nests, but people seem to really love plaiting them and I have got into the habit of doing it. Once you have done it a few times, plaiting a wigwam is quick but somewhat mad and you must do it all in one go.

You start at the bottom of the branches, at any point, then grab several branches at a time and twist them all together, creating a thick twig of a mane. Go upward with these but at a good angle, then go around and around the wigwam up and up, pulling in and plaiting all the loose branches from the birches' main trunks – think helter-skelter the whole time and move the dolly tub beforehand so that you can plod around its circumference fully as you go around and around. When you reach the tied top of the wigwam, you stuff the loose end into the pointed turret – often it needs to be tied in so that it doesn't ping out.

RIGHT Sweet peas climbing birch. The twiggy nature of the birch is a natural support for their touch-sensitive tendrils.

Afterword

Stop moaning and grow flowers for a greater, imaginary good would be an apt summary perhaps of this book.

How we garden, how we choose to grow and bring plants into our lives has never been more important than now as our planet faces a manmade, accelerating extinction of flora and fauna across the globe. Governments are not going to get us to where we need to be in properly fighting this shameful and heartbreaking loss of biodiversity or the disconnection that generations have been allowed to form toward nature; healing arguably will start and is starting in gardening attitudes. This book does not have all the answers, nor am I saying that my attitude is the right one to have, but I hope it is a bit of a help. My mind changes suddenly like the now climate-changed, drugged seasons do. I'm either in bloom with my petals fanned out or I'm dropping them with my head bowed down. What I do know are the smiles that even a window box of growing flowers can provide. Gardening won't save you but it can make you get up each day; nurturing is the best medicine.

With more people living in towns and cities, a human battery-farmed landscape, it is up to us as individuals to bridge this gap and encourage our families, friends and communities to see our gardens differently, as being more than just places for bins or stamps of mown grass. In an age where the world's rainforests are disappearing at a faster rate than when I was born, less than 30 years ago, we must have hope that gardens, even if they are measured in feet, can become the places that nurture our souls, through growing the faces of many beautiful flowers to see and adore every day. We can grow flower meadows and flower yards that will in turn create smiles and contentment that fuel divine bird song and ensure the buzzing of healthy bees, mini Edens that are more vital than ever – this, surely, is worth striving for.

The photos, which I find easier taking than writing, have been taken over the past three years. You can see I'm not a good garden photographer because I'm rarely up at first light. At the weekend, the hens get fed and let out with me in my dressing gown then I like to go back to bed! My camera is a very much abused Canon, EOS 1200D (I think that's right). I'd check but I don't know where it is – either in a nesting box of the hen house or in a bag for life somewhere, with its battery flat as a pancake and a drying-out tangerine for company, probably. I'd better go and have a forage. On that note, thank you for buying and reading my book; I hope that you have a life that is rich or becomes richer in flowers and living things.

Arthur Parkinson, 2021

Acknowledgements

Thank you to my mum, Jill, for putting up and supporting me and allowing/coping with this addiction, having boxes of bulbs under the kitchen table, chicken shit on the bricks, and watering the garden when I have been away. With thanks to my dad, Nick, for good kicks and telling me to get that bloody book written. Grandma Sheila for your endlessly positive and remarkable spirit. My friend, idol and saviour, Sarah, who has always been patient with me, and Caroline who I imagine is still ringing for a card machine helpline! Josie, Anita and Mary for lifts to and from Perch to the station. Jonathan, for putting up with me being around your lenses. Gorgeous Jorge and David and beautiful Juliet. JJM, Mr Tod, whose copper pot and garden will hopefully be brimming with hundreds, possibly thousands of awakening bulb beaks by the time you read this, upon your golden pheasant chair. Zoe and Amanda, Pauline and Aaron for friendship, Joanna for making me smile on the darkest of days, Adam for wise words and laughter. Alison and the team at Paradise Park for the flamingo photo. Marj for trips to Chatsworth. Thank you to Rachel, my agent, and Judith at Kyle. Lastly, to everyone in Instagram land who asked for this book and likes my photographs, cheers and thanks a lot sweeties.

LEFT Bumblebee enjoying the nectar and pollen of the annual cornflower 'Black Ball'. Her pollen baskets are loaded with its dark pollen.

Mail-order suppliers

UNITED KINGDOM

For mail-order plants, seeds and bulbs that are neonicotinoid free:

Sarahraven.com – autumn and summer bulbs, annual and perennial seeds, plug plants, vases, soil fertilisers and garden sundries.

Peternyssen.com – autumn and summer bulbs.

Claireaustin-hardyplants.co.uk – a generous range of hardy perennials.

Visiteaston.co.uk – a large range of sweet pea seeds harvested on site.

Davidaustinroses.com – bare-root and potted roses.
Plantsofdistinction.co.uk – new, rare and unusual flower and vegetable seeds.

Mail-order peat-free compost and plant feeds:
Naturalgrower.co.uk

Dalefootcomposts.co.uk

Garden hen houses, delivered and erected:
Domesticfowltrust.co.uk – attractive but practical and sturdy garden coops, also equipment and sundries.

Pest proof wild bird and chicken feeders:
Roam-wild.com

Vintage dolly tubs and teracotta pots:
Brownrigg-interiors.co.uk

UNITED STATES OF AMERICA

Ardeliafarm.com – sweet pea seeds.

Whiteflowerfarm.com – autumn bulbs.

Bearcreekfarm.com – dahlia tubers.

Selectseeds.com – annual flower seeds.

AUSTRALIA

Diggers.com.au – annual seeds.

Tesselaar.net.au – dahlias and tulips.

Lambley.com.au – bulbs, perennials and seeds.

Further reading

The Garden Jungle by Dave Goulson
(Vintage, 2020)

The Bold and Brilliant Garden by Sarah Raven
(Frances Lincoln, 2001)

Poultry for Anyone by Victoria Roberts
(Whittet Books Ltd, 1998)

Going Organic by Bob Flowerdew
(Kyle Cathie Limited, 2007)

The Flower Fix: Modern Arrangements for a Daily Dose of Nature by Anna Potter of Swallows and Damsons (White Lion Publishing, 2019)

The Garden at Chatsworth by Deborah Devonshire, Duchess of Devonshire (Frances Lincoln, 2001)

Instagram

Good and helpful accounts to follow for gardening advice and inspiration:

@pesticideactionnetworkuk – charity tackling the problems caused by pesticides.

@eppinggoodhoney – artisan bee keepers.

@myrealgarden – hugely helpful garden diary of Ann-Marie Powell.

@Francinehens – the hen keeper and writer Francine Raymond's garden.

@Beckycrowley_ – cut-flower grower at Chatsworth House, Derbyshire.

@Glennfacer – vegetable gardener at Chatsworth House, Derbyshire.

@Cotonmanorgarden – lots of pots and flamingos too!

@robinhhlucas – cottage gardener and beautiful arrangements.

@ewgardens – sweet peas.

@thelandgardeners – circus dahlias and making compost.

@perchhillgarden – a helpful diary by Sarah Raven's head gardener, Josie Lewis.

@mccormickcharlie – prince of dahlias and show vegetables.

@sarahravenperchhill – queen of dahlias.

@paradiseparkcornwall – a richly planted zoological bird garden in Hayle, St Ives, Cornwall.

@thecountrycrib – cottage gardeners.

@mikepalmer01 – a useful garden diary and interviewer of gardeners.

Arthur Parkinson's photography can be viewed on his Instagram account @arthurparkinson_

Index

Page numbers in *italics* refer to illustrations

UK/US glossary

allotment: community garden

car boot sale: flea market (also in Australia and South Africa)

charity shop: thrift store

dustbin: trash can

hundreds and thousands: sprinkles

lolly: popsicle

pavement: sidewalk

scrumping: stealing

IMAGES ON OPENING PAGES

PAGE 1 Buff Frizzle Pekin bantam hens, Patsy and Fergie, pecking in the yard at full autumn bloom. Plants from bottom left to bottom right: *Agastache aurantiaca* 'Navajo Sunset', *Tagetes linnaeus* growing with dill 'Tetra'. Dahlia 'Bishop of Auckland', cosmos 'Rubenza', Dahlias 'Soulman' and 'Waltzing Mathilda', chocolate cosmos and the cornflower 'Black Ball'. The purple fronds of the millet 'Red Jewel' are throughout.

PAGE 2 Tulip Bastia, the muppet tulip.

PAGE 4 Frizzle Buff Pekin hen, Fergie, amongst tulips 'La Belle Époque', 'Purple Dream', 'Queen of Night' and 'Antraciet'. These are short as they were not admittedly watered enough during the spring!

PAGES 6–7 Early summer. Coal buckets around the larger dolly tubs and bins will ensure a cascade of growth by the autumn. Left to right: sunflowers 'Earth Walker' and 'Claret', dahlias 'Totally Tangerine', 'Mango Madness', 'Schippers Bronze', 'Bishop of Auckland', 'Waltzing Mathilda' and 'Black Jack'. The pumpkin 'Crown Prince' is growing up a supporting arch of woven birch. Through this planting are annual cosmos 'Rubenza' and thunbergia 'African Sunset'. The fronds are from emerging gladioli, panicum grass and millets.

An Hachette UK Company
www.hachette.co.uk

First published in Great Britain in 2021
by Kyle Books, an imprint of
Octopus Publishing Group Limited
Carmelite House
50 Victoria Embankment
London EC4Y 0DZ

ISBN: 978 0 85783 917 6
Text and photography copyright 2021
© Arthur Parkinson*
Design and layout copyright 2021
© Octopus Publishing Group Limited

Distributed in the US by Hachette Book Group,
1290 Avenue of the Americas,
4th and 5th Floors, New York, NY 10104

Distributed in Canada by Canadian Manda Group,
664 Annette St., Toronto, Ontario, Canada M6S 2C8

Arthur Parkinson is hereby identified as the author of this work in accordance with Section 77 of the Copyright, Designs and Patents Act 1988.

Editorial Director: Judith Hannam
Publisher: Joanna Copestick
Assistant Editor: Florence Filose
Design: Miranda Harvey
Production: Nic Jones

*Photograph on page 76 © Jonathan Buckley
*Photograph on page 83 © Alison Hales

A Cataloguing in Publication record for this title is available from the British Library.
Printed and bound by Printer Trento S.r.l.

10 9 8 7 6 5 4